T0266826

KANSEI/AFFECTIVE
ENGINEERING

Industrial Innovation Series

Series Editor

Adedeji B. Badiru

Department of Systems and Engineering Management
Air Force Institute of Technology (AFIT) – Dayton, Ohio

Industrial Control Systems: Mathematical and Statistical Models and Techniques
Adedeji B. Badiru, Oye Ibidapo-Obe, & Babatunde J. Ayeni

Learning Curves: Theory, Models, and Applications
Mohamad Y. Jaber

Modern Construction: Productive and Lean Practices
Lincoln Harding Forbes

Project Management: Systems, Principles, and Applications
Adedeji B. Badiru

Statistical Techniques for Project Control
Adedeji B. Badiru

Technology Transfer and Commercialization of Environmental Remediation Technology
Mark N. Goltz

KANSEI/AFFECTIVE ENGINEERING

Edited by
MITSUO NAGAMACHI

CRC Press
Taylor & Francis Group
Boca Raton   London   New York

CRC Press is an imprint of the
Taylor & Francis Group, an **informa** business

CRC Press
Taylor & Francis Group
6000 Broken Sound Parkway NW, Suite 300
Boca Raton, FL 33487-2742

© 2011 by Taylor and Francis Group, LLC
CRC Press is an imprint of Taylor & Francis Group, an Informa business

No claim to original U.S. Government works

Printed in the United States of America on acid-free paper
10 9 8 7 6 5 4 3 2 1

International Standard Book Number: 978-1-4398-2133-6 (Paperback)

This book contains information obtained from authentic and highly regarded sources. Reasonable efforts have been made to publish reliable data and information, but the author and publisher cannot assume responsibility for the validity of all materials or the consequences of their use. The authors and publishers have attempted to trace the copyright holders of all material reproduced in this publication and apologize to copyright holders if permission to publish in this form has not been obtained. If any copyright material has not been acknowledged please write and let us know so we may rectify in any future reprint.

Except as permitted under U.S. Copyright Law, no part of this book may be reprinted, reproduced, transmitted, or utilized in any form by any electronic, mechanical, or other means, now known or hereafter invented, including photocopying, microfilming, and recording, or in any information storage or retrieval system, without written permission from the publishers.

For permission to photocopy or use material electronically from this work, please access www.copyright.com (http://www.copyright.com/) or contact the Copyright Clearance Center, Inc. (CCC), 222 Rosewood Drive, Danvers, MA 01923, 978-750-8400. CCC is a not-for-profit organization that provides licenses and registration for a variety of users. For organizations that have been granted a photocopy license by the CCC, a separate system of payment has been arranged.

Trademark Notice: Product or corporate names may be trademarks or registered trademarks, and are used only for identification and explanation without intent to infringe.

Library of Congress Cataloging-in-Publication Data

Nagamachi, Mitsuo, 1936-
  Kansei/affective engineering / editor, Mitsuo Nagamachi.
    p. cm. -- (Industrial innovation series ; Kansei engineering.)
  Includes bibliographical references and index.
  ISBN 978-1-4398-2133-6 (pbk. : alk. paper)
  1. Design--Human factors. 2. Human engineering. 3. System design--Psychological aspects. I. Title. II. Series.

TS170.N34 2011
658.5'75--dc22   2010030403

Visit the Taylor & Francis Web site at
http://www.taylorandfrancis.com

and the CRC Press Web site at
http://www.crcpress.com

Contents

Preface

The research of Kansei/affective engineering started in 1970 at Hiroshima University, and since then, more than 40 new Kansei products have been developed in Japan and worldwide. Those new Kansei products have been utilized in daily life. Today, Kansei/affective engineering has spread throughout the world. Many universities are teaching Kansei/affective engineering, and industries are using Kansei/affective engineering in innovative product development.

After coming back to Hiroshima from the University of Michigan, Ann Arbor, I worked as an ergonomist in vehicle design. I then worked as a consultant for Japanese automotive, steel plant, ship building, and many other companies as a manufacturing and quality control engineer. I noticed these companies had not produced products on the basis of the customer-oriented view. My thinking had been in human-oriented manufacturing, quality control, and management. People know me as the founder of the cell production system in Japan, in which only one worker assembles whole parts for a vehicle.

The Japanese term *kansei* means wants, needs, affect, emotion, and so forth. The concern of Kansei is the feeling that people have in their minds. If a customer feels a bit hungry he selects a restaurant that can serve a small meal. But if he wants a splendid dinner, he visits a high-class restaurant. In selecting a passenger car, the customer follows her wants, feelings, and motivation, while thinking of the price of the cars. Today all customers wish to purchase a product that matches their feeling (Kansei). In these recent decades of very severe economics in the world, the company that will survive is the one able to determine such a sense about customer wants, needs, and emotions—namely, Kansei.

Kansei/affective engineering has contributed to developing a lot of new Kansei products. For example, Sharp's new refrigerator, Sharp's Liquid Crystal Viewcam, Mazda's MX-5 sports car, Wacoal's Good-Up Bra, Komatsu's Ellesse (a shovel car), Milbon's Deesse's (a shampoo and treatment), BT's Lift car (Sweden), Panasonic Electric Works' Twin lamp (eco-lamp), and Rakmatair (a new mattress that prevents pressure sores), and many other products have been developed using Kansei/affective engineering. We have conducted research on the soft computing Kansei system, the computerized Kansei system for making an intelligent and virtual design based on the databases of customer emotion. We constructed the artificial Kansei system and neural network Kansei system, which supported the construction of design based on customer emotion. Kansei/affective engineering helps enhance workers' job satisfaction by considering their emotions. Kansei/affective engineering is an excellent technology that helps develop splendid and emotion-based products that match customer desire. As the new product is fit to the customers' feelings, the company makes a big profit from product sales.

We have aimed to promote this innovative technology worldwide for anyone to learn and apply to any kind of industry to develop new emotion-based products. First, the Kansei/affective engineer should observe the customer's behavior and determine his or her feelings, wants, and needs, or namely, emotions. Next, the engineer should have knowledge of statistical methods that can lead to good specifications for new product design. It is also very important that the engineer learns the human factors or ergonomics discipline, because all kinds of products should be easy for the customer to operate and use. Every product should be safe to use. The Kansei/affective engineer should also have a human orientation. All countries are going to become aging societies; the engineer should focus on the elderly and small children as well. Kansei/affective engineering needs to have a spirit of integration with a universal design philosophy.

This book is the product of the Nagamachi group of Kansei/affective engineers. Professors Tatsuo Nishino, Shigekazu Ishihara, Keiko Ishihara, Yokihiro Matsubara, Toshio Tsuchiya, Dr. Anitawati Mohd Lokman, and Dr. Ricardo Hirata Okamoto were my students; they have supported my Kansei/affective engineering research for a long time. Tom Childs and Jörgen Eklund have kindly collaborated for this book, contributing their Kansei/affective engineering research. I am very grateful for their efforts.

Finally, I would say that Kansei/affective engineering aims to realize three *wins*: Win for the customers in providing emotion-based products, Win for workers for satisfaction with their work, and Win for the company in achieving great profits from the Kansei products.

Mitsuo Nagamachi, Ph.D., CPE
Professor Emeritus, Hiroshima University
Professor Emeritus, Hiroshima International University
Professor Emeritus, Kure National Institute of Technology

About the Editor

Mitsuo Nagamachi, Ph.D., is the founder of Kansei engineering/Kansei ergonomics, an ergonomic new product development technology known and implemented worldwide. As a professor at Hiroshima University, Dr. Nagamachi created more than 40 new Kansei products, including cars, construction machinery, home appliances, brassieres, cosmetic products, handrails, toilets, and even a bridge over a river.

Dr. Nagamachi received his Ph.D. in mathematical psychology from Hiroshima University in 1963. He then studied medicine and engineering. From 1967 to 1968 he was a guest scientist at the Transportation Research Institute of the University of Michigan. Upon his return, he became the youngest ergonomic researcher appointed to Japan's Automotive Research Committee, whose mission was to make the Japanese automotive industry a world player. Dr. Nagamachi has consulted with the Japanese automotive industry on manufacturing, quality control, vehicle safety, management robotics, and Kaizen. In the 1970s, he began his research on Kansei engineering, which translates consumer's psychological feelings about a product into perceptual design elements. This technique resulted in the creation of numerous phenomenally successful products, including the MX-5 for Mazda, the Liquid Crystal Viewcam for Sharp, and the Good Up Bra for Wacoal.

Dr. Nagamachi has traveled extensively to teach Kansei engineering. He had served as a consultant in England, Spain, Sweden, Finland, Mexico, Taiwan, Korea, and Malaysia. In 2008 he was awarded the Japan Government Prize for the founding of Kansei engineering. He has received many academic awards from the Japan Society of Kansei Engineering. He has published 89 books and 200 articles.

Contributors

Ebru Ayas
Department of Ergonomics
Royal Institute of Technology

Cathy Barnes
Faraday

Tom Childs
Professor Emeritus
Manufacturing Engineering
University of Leeds
United Kingdom

Jörgen Eklund
Department of Ergonomics
Royal Institute of Technology

Keiko Ishihara
Department of Communication
Hiroshima International University
Japan

Shigekazu Ishihara
Department of Kansei Design
Hiroshima International University
Japan

Stephen Lillford
Design Perspectives
United Kingdom

Anitawati Mohd Lokman
Faculty of Computer and
    Mathematical Sciences
Universiti Teknologi MARA (UiTM)
Malaysia

Yukihiro Matsubara
Faculty of Information Sciences
Hiroshima City University
Japan

Tatsuo Nishino
Department of Kansei Design
Hiroshima International University
Japan

Ricardo Hirata Okamoto
Keisen Consultores
Mexico

1

Kansei/Affective Engineering and History of Kansei/Affective Engineering in the World

Mitsuo Nagamachi

CONTENTS

1.1 What Is Kansei?

Imagine a scenario where you are searching for a restaurant during lunchtime. You are very hungry and find a restaurant you are not familiar with. When you enter, you first meet a waitress. She welcomes you and guides you to a table. You order a dish, and while you wait you look around the room. Then, you smell the aroma and are pleasantly surprised at the sight of the exquisite cuisine the server places on your table. The taste is beyond your expectations. Your impression of the restaurant escalates and makes you feel splendid.

When you first entered the restaurant and met the waitress, you felt some abstract feeling. When you looked around the interior, you had a good sense about the place. You felt pleased with the restaurant. Then, the cuisine was great. These feelings are all Kansei. *Kansei* is a Japanese word that expresses the feelings gathered through sight, hearing, smell, and taste. In our scenario, finally you think of this restaurant as splendid and someday you want to take your family there. This is also Kansei.

Imagine now another scenario where a woman goes shopping but has no specific thing to buy. She walks around in a department store and finds a medium-sized handbag at a low price. She is fascinated with it, especially with its color. This fascination is also a kind of Kansei.

Kansei is a Japanese term with a broad interpretation, including

1. Sense, sensitivity, sensitiveness, sensibility
2. Feeling, image, affection, emotion, want, need

Consider a man who has keen senses when he notices the events around him. In this case, we say he has good Kansei. Also, if a man is able to relate to children and animals, we say he has the Kansei (the sense). When a manager teaches his subordinates, he says you should have Kansei about customers, which means that customer service personnel should try to understand customers' feelings. The term *Kansei* has such wide meanings, and in this book we have used the term as it is because there is no accurate translation in other languages, particularly in English.

The Kansei of Kansei/affective engineering applies mainly to the customers' feeling. If research and development (R&D) people are oriented to the customers' wants and needs, the team will be successful in developing a good product, and the customer service people can fulfill the customers' expectation. The service is also one of products, namely the *service product*. There are two different streams in product development, which are called *product out* and *market in*. The former implies a philosophy of product development based on technology developed in a company or based on the company strategy, *without* attention to customers' wants and needs. Many inventions have emerged from this approach. Another approach to product development is to focus on customer wants and needs. Nowadays people have many goods at home, and it is not easy to stimulate their purchasing behavior. But customer-oriented product development will be successful in selling a new product because the market-in philosophy leads to the development of a product that fits customers' feelings and emotions. This is why Kansei-oriented development is needed in R&D activities.

On the other hand, Kansei/affective engineering is oriented to human minds. This is why it is called *human-oriented* product development. The first target of Kansei/affective engineering is to grasp human Kansei, and then if new technology is needed, Kansei engineering will seek the new technology development in order to realize the Kansei product.

1.2 What Is Kansei/Affective Engineering?

Kansei engineering is a kind of technology that translates the customer's feeling into design specifications (Nagamachi and Lokman 2010). The R&D team grasps the customer's feeling, namely the Kansei; analyzes the Kansei data using psychological, ergonomic, medical, or engineering methods; and designs the new product based on the analyzed information. Kansei/affective engineering is a technological and engineering process from Kansei data to design specifications.

People's lives are diverse, but fundamentally all people seek pleasant and emotional satisfaction in quality of life (QOL). It is becoming important to determine the satisfaction people have in mind that will enhance their QOL. On the other hand, people are very aware of the ecosystem. Air, water, and temperature are becoming more integral parts of people's lives. In addition, most countries are becoming older societies, and the welfare field is another new issue to address. This multifaceted consciousness is also included in Kansei. These issues should be considered during product development as a Kansei ecosystem.

The process of Kansei/affective engineering should include the following scheme: First, a Kansei engineer should think, Who are the customers? Second, What do they want and need?; that is, what is their Kansei? Third, the Kansei engineer should consider how to evaluate the customers' Kansei. After the Kansei evaluation, the engineer should analyze the Kansei data using statistical analysis or psychophysiological measurement, and then transfer the analyzed data to the design domain.

1.3 Routes to Reach a New Kansei Product

1.3.1 Psychological Phase of the Kansei

The Kansei is an outcome through cognition and the five senses: sight, hearing, taste, smell, and touch. The inner sense is related to gravity, and it is useful to test the feeling as in when speeding up or slowing down a vehicle. Accordingly, to be more precise, we have six senses. There is the cognition function, which is concerned with memory, judgment, interpretation, and thinking. The Kansei comes out through cognition after some work by the senses. In our earlier story of a new restaurant, a customer meets a waitress and hears her voice. The cuisine is served, and the customer smells and tastes the food as he eats. These are sensations, and the customer feels that the restaurant is friendly and warm. These are Kansei that emerged through cognition with sensation activities.

When you want to make a new, good Kansei product, you should first think what Kansei are related to the new product and how to obtain the customer's Kansei. In a restaurant business, the owner should think of what factors stimulate a customer to develop his/her feelings and motivation. Are these due to friendly service, or the decor of the room, or a good cook's cuisine? Of course, price is also a concern.

1.3.2 Psychophysiological Phase of the Kansei

The friendly voice is related to the physiological mechanism of the ear, and perceiving a good taste is a physiological function of the tongue. Hard work forces workers to exhaust their energy and makes them tired. Brain waves (EEG) are stimulated when working with high motivation, but idle and repetitive work increases the worker's feeling of boredom. When using a very soft mattress, people could feel uncomfortable if they have high body pressure. These are examples of a kind of Kansei known as psychophysiological Kansei.

1.3.3 Routes to Reach the Kansei

The customer's Kansei has a diversity of expressions, from psychological to psychophysiological measurement, and each measure also has a variety of emergence, as shown in Figure 1.1. The Kansei engineer who wants to make a new Kansei product should first choose the most appropriate route to reach the correct customer Kansei, by the use of EEG, EMG, attitudes, or words.

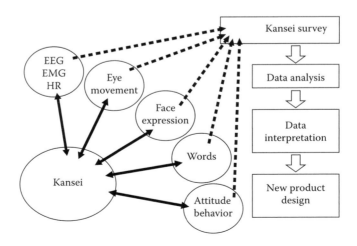

FIGURE 1.1
Choices of route to reach the Kansei.

This choice is important. If you choose the right route, you will be successful in achieving a Kansei product design. But if you cannot choose a route that reaches the correct Kansei, you cannot successfully perform Kansei engineering. The Kansei engineer should first observe the customer's behavior and check which route will best reach the customer's Kansei. The successful route is not always a single one. It can be a combination of several routes.

1.4 What Is Designing Based on Kansei/Affective Engineering?

Kansei/affective engineering is defined as the technology of translating the consumer's Kansei into the product design domain (Nagamachi 1995, 1999, 2005, 2010). The process of performing Kansei/affective engineering is shown in Figure 1.1:

1. Grasp the consumer's Kansei in the specific product domain (passenger car, cosmetic, shaver, etc.) using psychological or psychophysiological measurements.
2. Analyze the Kansei data by statistical, medical, or engineering methods in order to clarify the Kansei structure.
3. Interpret the analyzed data and transfer the data to the new product domain.
4. Finally, design a new Kansei product.

Following this Kansei/affective engineering process ensures you will get to the fourth stage automatically, but this does not always produce successful product development. Statistical analysis can make clear the Kansei structure, but this does not go beyond the existing data level. To reach inventive and innovative product development requires a Kansei engineer and a product designer to collaborate and milk their idea for producing an excellent Kansei product using the Kansei data. The process is illustrated in Figure 1.2.

As described in the next paragraphs, we have developed a variety of Kansei/affective methods, from category classification to an artificial intelligence system. We describe here two very simple examples of Kansei product development, which can be applied by anyone to develop a Kansei product easily. In any case, Kansei product development must focus on customer-oriented or human-oriented aspects.

The Sharp Company introduced Kansei/affective engineering into its design group and used it to develop a new refrigerator in 1978. The project team, supported by this author, visited monitors' houses with a camcorder in order to observe how they use a refrigerator. The team set up the camcorder in front of the refrigerator and took pictures of a woman operating it.

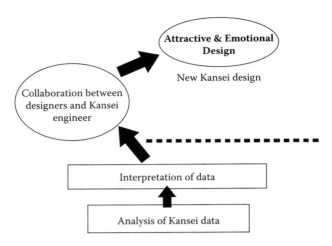

FIGURE 1.2
Sublimating process of a Kansei product development.

After that, the team checked the images and found that about 70% of pictures were of the woman opening the lower door and bending very frequently as she picked out vegetables to cook. From an ergonomic viewpoint, energy used for a bent posture is three times as much as for a standing posture (Nagamachi 1996). An illustration of this is shown in Figure 1.3. Although

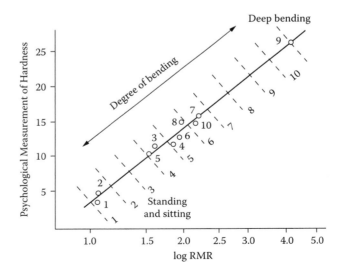

FIGURE 1.3
Relationship between bent posture and body energy.

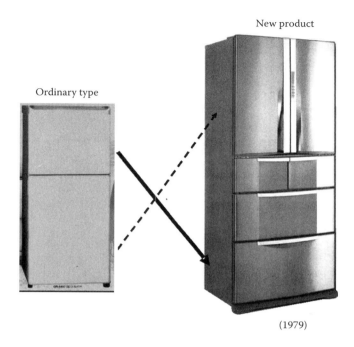

New product

Ordinary type

(1979)

FIGURE 1.4
The new refrigerator (right) developed from an ergonomic survey.

the customer did not complain about her bent posture, an ergonomist suggested improving this situation.

After this author suggested that the posture aspect is one element of the customer's needs, the team proposed that the freezer on the top should be moved to the bottom and the vegetable drawer should be put on top in order to promote easy operation, as shown in Figure 1.4. The new Kansei refrigerator was developed in 1979 by Sharp, and this new type with several drawers became the Japanese standard. Sharp produced various refrigerators ordered from competitors as the original equipment manufacturer and became the number one manufacturer of refrigerators at that time. If a Kansei engineer has a good sense about the customer's needs, he/she can create an innovative product like the one in this story.

Another R&D team at Sharp created a new camera, the Liquid Crystal Viewcam. Sharp manufactured the camcorder, but its market share was just 3%. The CEO wanted to cease production of the camcorder. However, inspired by the success of the new refrigerator, the camcorder team appealed to the CEO to continue the development of a new product based on Kansei engineering. The team followed the same process as in the refrigerator's development. They visited subjects' homes to take pictures of camcorder usage situations and found the same bent posture problem as in the refrigerator

FIGURE 1.5
The Liquid Crystal Viewcam created by Sharp.

research. The Kansei team discussed and reached a Kansei idea, to rotate a lens 350°, and to check an image by a liquid crystal mirror behind the camera instantly after taking picture. The Liquid Crystal Viewcam developed in 1980 was extended to the creation of a digital camera, and nowadays everybody enjoys taking pictures with the product. The Liquid Crystal Viewcam was selected for the Good Design Award by the Japanese government.

The stories of these two new product developments show us how Kansei/ affective engineering can lead to a new product invention, and if we have a Kansei hint and sense related to the customers' needs, we can easily produce new inventions.

1.5 History of Kansei/Affective Engineering

1.5.1 History of the New Products Developed Using Kansei/Affective Engineering

This author is basically a psychologist. After obtaining a Ph.D. from Hiroshima University, I began providing consultation services to manufacturing companies, quality control management (called TQC, or total quality control, in Japan), and safety management. I am also an industrial engineer. I have consulted for many Japanese enterprises, such as Nissan, Toyota, Honda, and Mitsubishi Motors, as well as Nippon Steel, Matsushita, Matsushita Electric Works, Kubota, and many others. My clients, all large enterprises, account for more than 300 companies. During my consultation activities, I had the idea that people are expecting new types of products that will fit their sensations and emotions. I began Kansei/affective engineering research in 1970.

In the beginning, in collaboration with Japan's IBM and NEC, I built an expert system that was able to suggest good fashion design when Kansei words such as *beautiful, elegant,* and so forth are input into a computer. The system, called FAIMS (Fashion Image Expert System), was very successful in determining all kinds of female students' fashions. Following this research, I constructed many Kansei expert systems such as HULIS (house design), ViVA (kitchen design based on Kansei virtual technology), Cockpit (cockpit room design of construction machines), GAINT (vehicle interior design based on GA), WIDIAS (word image diagnosis Kansei system), HousMall (house designing system), and others (Nagamachi et al. 1974, 1988; Nagamachi 1977; Nomura et al. 1998). Some of the computerized intelligent software has been utilized in several enterprises.

In addition to Sharp's refrigerator and Liquid Crystal Viewcam development, I have also endeavored to apply the application to businesses. First, the CEO of Mazda, Mr. Kenichi Yamamoto, asked me to help with an introduction of Kansei/affective engineering into R&D on its next sports car. I taught the project team how to implement Kansei/affective engineering. Because the project was related to a sports car for young drivers, a team member sat in the passenger seat and took pictures of the driver's maneuvers with a camcorder. Another team member stood at an intersection and took pictures of the car maneuvering when it came through an intersection. These pictures were analyzed frame by frame, and the team members wrote keywords on a piece of paper. This procedure is called the *card method*. About 600 pieces of paper were produced, and these were constructed into a tree structure, which is the *category classification method* (CCM). The CCM suggested many design elements, each of which was examined through ergonomic experiments. Finally, these elements were integrated into the Mazda MX5 (Miata). Kansei/affective engineering was applied to the engine innovation, interior and exterior design, and, overall, into the entire design of the MX5. The MX5 is still very popular all over the world today, even though the original dates back to 1987.

Komatsu, which is a construction machine maker, asked me to create a modern design for a shovel car, using Kansei/affective engineering. We conducted a survey of shovel car drivers. Based on the drivers' emotional data, we designed quite a new style and a much more advanced model, compared with the original square, sand-colored machine. The new design of Komatsu's shovel car (Figure 1.6) had a beautiful and advanced shape and was available in purple. The new design changed the paradigm for all auto makers. It received the Japanese government's Good Design Award.

In 1979, Matsushita Electric Works (now Panasonic Electric Works) asked me to do research to find a lighting system that gives people a relaxed feeling. I constructed a guest room inside the company's office, set up different kinds of

FIGURE 1.6
Komatsu's Kansei design of a shovel car (PC50).

lighting systems inside it, and measured people's feelings of relaxation using an experimental approach. The company's designer group spent time in this experimental room and completed the questionnaire about their feelings on each lighting condition. The ergonomic experiment took a long time to reach its conclusions. The conclusions for relaxing lighting were (1) the illumination level should be 300–400 lx; (2) a cooler temperature is recommended around 3700 K; and (3) a fluorescent lamp is hard pressed to satisfy these conditions. The interpretation of the ergonomic experiment suggested that the fluorescent lamp should be bent into two folds, and if possible a fluorescent lamp with low color temperature will be in a small bulb like a tungsten bulb.

Matsushita made a bent two-folds and four-folds fluorescent lamp, and finally it was successful in producing a fluorescent bulb. As the first developed product was a bent fluorescent lamp, these innovative products were called *twin lamps*. As a result, the electric expenditure was reduced to one-fifth compared with that used by the original product. The Japanese government decided to change all lighting systems to the new fluorescent lamp.

Kansei/affective engineering is also effective regarding the ecosystem. Kansei/affective research is called Kansei ergonomics, if the product development was conducted based on ergonomic experimentation.

More than 40 new types of products have been created by this author using Kansei/affective engineering. The description of other new Kansei products will be continued in Chapter 2.

FIGURE 1.7
Kansei/affective ergonomics created a reduced energy lamp called a twin lamp.

1.5.2 History of Kansei /Affective Engineering Research

In 1970, I started my research on Kansei engineering at the University of Hiroshima. I concentrated first on the construction of an intelligent computerized system with the cooperation of my graduate students. FAIMS and HULIS are the results of that research. My group has addressed Kansei engineering research in many international conferences, especially at ODAM (the conference of Organizational Design and Management) for 10 years. With the support of Dr. Soon Yo Lee of Korea University, I have taught Kansei engineering at many universities in Korea. I also cooperated with Dr. Lee in relation to Japan and Korea collaborating on Kansei engineering, and they founded the Japan-Korea Conference of Kansei Engineering. These activities influenced and prevailed in Kansei engineering research in Korea, where Korean research then established the Korean Kansei Engineering Society in 1997. I initiated Japanese researchers, and the Japanese Society of Kansei Engineering was established in 1998.

As a result of joint research with Professor Jörgen Eklund of Linköping University in Sweden, an exchange program concerning research on Kansei engineering was established via the Internet. Both groups planned collaborative research in which students at both universities communicate through the Internet, and they have constructed a new product using the Kansei database developed in Japan (Nagamachi 1998). Later, Professor Tom Childs of the University of Leeds joined our research group, and he and his colleagues started Kansei design research. I2BC (International Institute of Human Well-being), which is the third sector of Spain's Andalucía Government, introduced Kansei/affective engineering as a new division with my help in 2009. It aims to build a European Excellent Center of Kansei/Affective Engineering.

In Asia, along with Japan and Korea, Malaysia is yet another country with interest in Kansei/affective engineering. Universiti Teknologi MARA is one such group, and Dr. Anitawati Mohd Lokman is planning to introduce a Kansei design course in her department. There are many researchers who show a lot of interest in Kansei/affective engineering nowadays, and the number keeps increasing. Taiwan researchers started Kansei/affective engineering, and a number of small enterprises have introduced it in their own companies.

In Mexico, Dr. Ricardo Hirarata Okamoto, who is a Kaizen specialist, is expanding Kansei/affective engineering in Mexico, and he is very active in consulting with many companies on Kansei engineering applications.

Kansei/affective engineering has spread all over the world, and we have developed a lot of new products that are results of the application of Kansei/affective engineering pertaining to emotions. It aims at customer-oriented or human-oriented product development and it is an advanced and innovative technology that will enhance people's QOL.

References

Nagamachi, M., Senuma, K., Iwashige, R. (1974). A research on emotional technology, *Jap. J. of Ergonomics Soc.*, 10 (2), 121–130.

Nagamachi, M. (1977). Emotional analysis on a room atmosphere, *Jap. J. of Ergonomics Soc.*, 13 (1), 7–14.

Nagamachi, M., Ito, K., Tsuji, T., Chino, T. (1988). A study of costume design consultation system based on knowledge engineering, *Jap. J. of Ergonomics Soc.*, 24 (5), 281–289.

Nagamachi, M. (1995). Kansei engineering: A new ergonomic consumer-oriented technology for product development, *International Journal of Industrial Ergonomics*, 15 (1), 3–11.

Nagamachi, M. (1996). *Ergofactory: Challenge to Comfortable Factory*, Japan Plant Maintenance Association, Tokyo.

Nagamachi, M. (1998). Kansei designing group work system through Internet, Manufacturing and Hybrid Automation-II, 63–66.

Nagamachi, M. (1999). Kansei engineering: A new consumer-oriented technology for product development, in W. Karwowski and W. S. Marras (Eds), *The occupational ergonomics*, Chapter 102, 1835–1848.

Nagamachi, M. (2005). Kansei engineering, in N. Stanton, A. Hedge, K. Brookhuis, E. Salas, and H. Hendrick (Eds.), *Handbook of human factors and ergonomics methods*, Chapter 83, 83-1–83-4. CRC Press, New York.

Nagamachi, M., and Lokman, A. (2010). *Innovation for Kansei/affective engineering*, CRC Press (in press).

Nomura, J., Imamura, N., Enomoto, N., and Nagamachi, M. (1998). Virtual space decision support system using Kansei engineering, in T. Kunii and A. Luciani, (Eds.) *Cyberworlds*, Chapter 18, 273–288, Springer, Tokyo.

2

Methods of Kansei/Affective Engineering and Specific Cases of Kansei Products

Mitsuo Nagamachi

CONTENTS

2.1 Category Classification

In this chapter, methods of Kansei/affective engineering that have been developed to date will be described together with an illustrative case of Kansei product development.

Category classification is a tree structure from a main event to subsequent subevents, as shown in Table 2.1. This method was utilized by Mazda. The CEO decided on a new car development in which the target was young drivers and the product domain was a sports car. The project team sat next to the driver with a camcorder and recorded the driver's operation. Another team stood in an intersection and recorded the young driver's maneuvering. After that, the team members examined all of the picture frames and used the *card method* to record their findings. When they got a hint or suggestion from pictures, they noted a keyword on each small card (called a K-card); one word on one card. If the team found several K-cards with a similar concepts or meanings, these cards were gathered into one group. The total number of K-cards from examining the pictures was about 600, which were organized into about 20 groups, which is an average of 30 cards in each group. These

TABLE 2.1

Category Classification Utilized in Miata Development

Kansei				Physical	Ergonomic	Automotive
Zero	1st	2nd	nth	Traits	Experiment	Engineering
	-- Tight		· · · ·	Size	Tight feeling	Chassis design
	Feeling		· · · ·	Width	experiment	Seat design
	-- Direct		· · · ·	Height	Interior Kansei	Interior design
OHM	Feeling		· · · ·	Seat	experiment	Power train
	-- Speedy		· · · ·	Steering	Steering	development
	Feeling		· · · ·	Shift lever	function test	Steering design
	--Communi-		· · · ·	Speedometer	Noise frequency	Speedometer
	cation		· · · ·	Open style	analysis	design

groups were arranged from top concepts to more fundamental groups in a tree structure, as shown in Table 2.1 (Nagamachi 1995).

In developing the Mazda Miata, about 600 keyword cards were classified into about 20 groups, and then these groups were rearranged from the top-level concept to sublevels. All card groups were represented by a name, and finally a top concept was named "one-human-machine" (or unification of human and machine), which implies that the young drivers need an emotional feeling or want a unified connection between human emotion and machine movement and/or function. Category classification means that there is a tree structure from a top concept through subconcepts. The subconcepts in the nth level were transferred to the ergonomic experiment phase, in which the ergonomic experiments produced the detailed specifications of the nth level. The analyzed details were transferred finally to the design domain. In this case, the details of design specifications were integrated into an automotive design of a sports car.

The concept classification method is easy to perform. Anyone can learn the process very easily. Today, we try this method first for all kinds of product development. For this process, we observe target client behavior and try to conduct a psychological survey of the clients. After analysis of the observed data we consider what the clients want and what kind of emotional feeling they have. From this research we decide the top concept of the product development and then we start the category classification survey.

We interpret the meaning of the top concept and put down a keyword on K-cards in the first level. In the second level, we deploy further each concept to the subconcepts, and this continues to the nth level. Subsequently, these subconcepts in the nth level become the real data through the ergonomic survey and are transferred to the design domain. In general, the deployment of category levels stops around the third or the fourth level. In the case of the Miata (MX5), the project team first carried out the concept deployment work in order to build a new sports car image for a young driver, and then each of the final subconcepts was transferred to the ergonomic experiments. For instance, a steering gear ergonomic experiment was conducted using a variety of experimental gears in different lengths and with different torques. Then an experiment was conducted to decide the length and torque, performed using the company staff. The subjects touched and moved the different steering gears and scored their feelings on a 10-point scale, with the best fit of the emotional feeling "I am controlling this machine." At the conclusion of the ergonomic experiment, it was found that 9.5cm is the best fit to the self-controlling emotion, and this length of the steering gear was implemented in the final design of the Miata. All Miata parts, from the engine, exterior, and interior, to the seats and the steering wheel, were decided using a similar process. The Miata became very popular all over the world, due to its aesthetic, structural, and functional design, all decided through the emotional research using Kansei/affective engineering.

Another case of new product development is in a shampoo and hair treatment product produced by Milbon, which is called Deesse's. This will be described later because it used the Kansei/affective method that combined the category classification and Kansei engineering Type I.

2.2 Kansei Engineering Type I

2.2.1 Method of Kansei/Affective Engineering Type I

Kansei engineering Type I is a fundamental technique of the Kansei/affective engineering method which uses the process-ruled means. Everybody can follow the ruled process to reach the final successful conclusion (Nagamachi and Lokman 2010).

TABLE 2.2

The Process of Kansei/Affective Engineering Type I

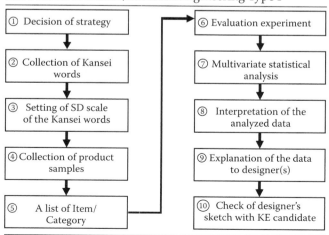

Kansei/affective engineering Type I has 10 steps:

1. Decision of strategy. A Kansei engineer listens to the client company's CEO or top R&D manager and understands the company's new product development strategy. The most important point for a Kansei/affective engineer is to grasp what kind of requirements the company has and what will give them the highest satisfaction in new product development.

2. Collection of Kansei words. After understanding the client company's strategy, the Kansei engineer collects Kansei words related to the product domain. We usually synthesize from related magazines, business newspapers, or salespeople's information concerning customer emotion and opinion. The Kansei words are adjectives, nouns, or verbs, and sometimes sentences. *Beautiful, elegant, premium, smart, simple, large, colorful, red, blue, square, easy to open*, and so forth are all Kansei words. It is recommended that you first collect a lot of Kansei words and then reduce these to a small number of very important and relevant words.

3. Develop an SD scale. The SD scale (the *semantic differential*) is a psychological measurement scale devised by C. E. Osgood and his colleagues (Osgood et al. 1957). This method is used to make clear the psychological language structure. Osgood arranged positive and negative words on both sides of a horizontal line. For instance, *beautiful—ugly* are set on both side of a continuum. But Kansei/affective engineering is intended to achieve a good design, not an ugly design. So, we arrange positive and negative Kansei words on both sides of the scale such as *beautiful—not beautiful*. There are several scales, 5-scale, 7-scale, 9-scale, and 11-scale, but the 5-scale is the easiest to understand and the easiest for clients to use.

4. Collection of a sample product. The Kansei/affective engineer should gather products that are similar to the targeted product. If the targeted product is a shampoo bottle, the engineer collects many similar shampoo bottles from the market. If it is an automotive exterior design of a passenger car, the engineer collects many passenger vehicles. About 20 or 25 samples are usually enough.

5. Make a list of *item/category*. Item/category is related to the final design specifications: *item* implies the design item of the sample product, and *category* means the detail of the design item. For instance, color, shape, size, roundness, and so forth are examples of items; and red, yellow, green, blue, and so forth are the categories for the color item. The Kansei/affective engineer should be very careful of the sample product's items and categories. A very refined classification of the items and categories will lead to a successful design.

6. Evaluation experiment. The sixth step is to conduct the evaluation experiment using subjects. The subjects receive an instruction and evaluate each sample with the 5-point SD scale of Kansei words.

7. Statistical analysis. The evaluated data are analyzed using a multivariate statistical analysis. In this method, we utilize correlation coefficients to check the relationship of meanings between Kansei words, principal component analysis (PCA) for positioning, factor analysis to make clear the sample data structure, and finally quantification theory Type I (QTI) or partial least squares (PLS) to identify the design element relevant to the specific emotion.

8. Interpretation of the analyzed data. Each statistical analysis has a specific interpretation property. Correlation coefficient implies the similarity in meaning between each Kansei word, and PCA is able to show us positioning interrelated among Kansei and sample products. Factor analysis shows us the psychological structure of Kansei words related to the selected product sphere and sample product position related to the Kansei structure. QTI or PLS tells us what Kansei words will have what kinds of design specifications. We interpret the data and integrate them into the product design properties.

9. Explanation of interpretation of the data to a designer. The most important step is the collaboration with a product designer. The Kansei/affective engineer should explain the analyzed data and the interpretation to the designer. Sometimes several suggestions are derived from data analysis. The engineer has to motivate and stimulate the designer to understand the final data interpretation and to draw out the designer's new design idea of emotional design beyond the data, as shown in Figure 1.2.

10. Check the new design idea. Finally, the Kansei engineer should evaluate whether the newly designed product will fit the customer's emotion and whether it reveals the emotional design. If not, she motivates the designer to a better intrinsic design idea.

2.2.2 Shampoo and Hair Treatment: Deesse's Development

Milbon, a cosmetic maker, asked Nagamachi to assist the company in making a new shampoo and hair treatment. First, we visited many salons to survey hair problems of 200 ladies, using a structured interview method. The data obtained were analyzed by quantification theory Type III (similar to factor analysis), and we concluded that what was needed was a kind of material that could hold their hairstyle even when a strong wind blows. We decided to use the category classification method, which Mazda had utilized for Miata development, and the *zero concept* was settled as *soft and rustling (breezy) hair*. We deployed the top concept to subconcepts as shown in Figure 2.2. In the third level of deployment, we stopped the concept deployment and selected

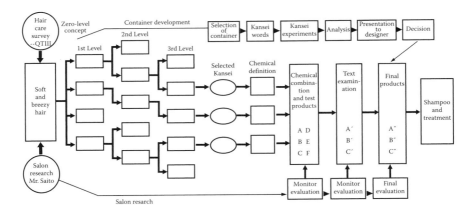

FIGURE 2.1
The product development process of a shampoo and treatment.

FIGURE 2.2
Milbon's shampoo and treatment, Deesse's.

the three most important subconcepts and conveyed them to the research institute to make 600 different materials combined with different properties. Using mannequins, we evaluated the newly developed materials and compared them with the ideal and conceptual property. Then, the 20 remaining new materials were evaluated using real subjects, and only two materials remained. Finally, the R&D team chose one new material.

This was the process of making the decision of shampoo and treatment bulk (chemical material). The next research was to decide the container shape and color. Here, we adopted Kansei/affective engineering Type I.

For a container design development, we collected 62 different containers of shampoo from the market and decided to use Kansei/affective engineering Type I. We arranged 20 Kansei words related to the containers' Kansei and asked 30 female subjects to evaluate these samples on the 5-point SD scale of Kansei words. The evaluation data were then analyzed using PCA and factor analysis. From the PCA data, we interpreted the relations between the *soft and rustling* Kansei and samples in its positioning analysis. From factor analysis we found the factors including *soft and rustling* Kansei. We obtained the design properties from QTI calculations related to the top concept. Finally, we collaborated with designers, and the Kansei engineer suggested the emotional design cues in order to go beyond the data level. The flow line from left to right implies the chemical bulk development using category classification, and the top line in the figure shows the container design development using Kansei/affective engineering Type I.

Figure 2.2 shows the final product design in the developmental process of the new shampoo and treatment. The left product is the shampoo and the right one is the treatment. Milbon was successful in selling these emotion-based products. The new product became very popular among both young and older women, and Milbon made a large profit from this one product (Nagamachi 2001).

2.2.3 Brassiere Good-Up Bra Development

Wacoal, a well-known lingerie maker, asked Nagamachi to introduce Kansei/affective engineering into its design division. It surveyed 2000 women about the emotional feeling when wearing a brassiere. They answered that they wanted to become *beautiful* and *graceful*. The R&D team collected a variety of different brassieres from different makers and invited 200 women to respond to a survey about their feeling. Each subject wore one of the sample brassieres and evaluated it on the SD scale of *beautiful* and *graceful*. Then, the team analyzed those sample products from the aspect of engineering and finally found the Kansei principle: The new brassiere design should be made so that two breasts would reside within two body lines, in parallel and pointing a little upward. If it is so, then the new brassiere will induce

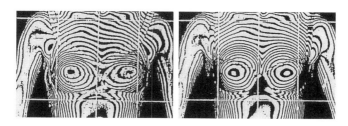

FIGURE 2.3
Moire analysis for ordinary brassiere (left) and the new brand (right).

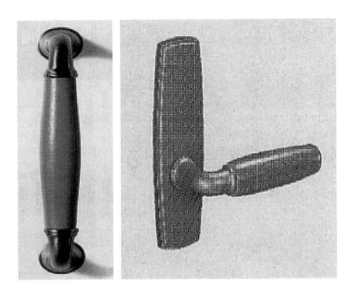

FIGURE 2.4
Premium and *easy in operation* doorknob.

a feeling of *beautiful* and *graceful*. The new brassiere was named Good-Up Bra, and it was a very big hit on the market. Many young women bought it and enjoyed wearing it.

2.2.4 A Doorknob Development

Another application of Kansei/affective engineering type I was to make a *premium* doorknob. We attempted to build in *very premium* and *easy in operation* doorknobs for all doors produced by Panasonic Electric Works. We collected 105 different doorknob samples and evaluated them on the 5-point SD scale of Kansei words. University students joined the evaluation experiment. We analyzed the evaluated data using PCA, factor analysis, and QTI, and we obtained a good conclusion. A Kansei/affective engineer and a designer collaborated and completed the final design of the premium doorknob based on the Kansei data as shown in Figure 2.4. Today, the new doorknob is attached to all doors produced by Panasonic Electric Works.

2.3 Kansei Ergonomics

All Kansei products are designed with the use of Kansei/affective engineering as well as the concept of ergonomics, especially concerning easy handling or operation. If we concentrate on ergonomics to enhance the Kansei product design and we need to implement the ergonomic idea into a Kansei product, we call it *Kansei ergonomics*. Accordingly, during Kansei product development, we perform ergonomic evaluation, or we conduct the implementation of ergonomic principles. The implementation of ergonomics into Kansei product development depends on the product property and ergonomic implications to enhance the customer's QOL (quality of life).

2.3.1 Designing a Toilet for Elderly People

Panasonic Electric Works wanted to create a new type of toilet, but Nagamachi suggested designing an easy and useful one for elderly people, because Japan is becoming an aged society. We decided to apply Kansei/ affective engineering Type I and collect 13 different types of toilet from different makers. We created the SD Kansei scale with keywords for when using the toilet, namely the keywords of emotional feelings when sitting on the toilet seat and when standing up. Young and old subjects participated in this research. The evaluation data were then analyzed using PCA, factor analysis and QTI. In using QTI, we had to arrange the item/ category data. For the latter work, we measured the three-dimensional features (width, shape, curve, etc.) of each sample. For Kansei keywords, we focused on *ease in sitting and standing up* and *premium sitting feeling*. We grasped the relations between these two Kansei words and a toilet property, and we obtained the ergonomic design principles to realize the emotional feeling.

As a result, we designed a new toilet surface that is easy to use by elderly people, with two arm rests and with the curved surface in a three-dimensional shape, as shown in Figure 2.5. In addition to this, we

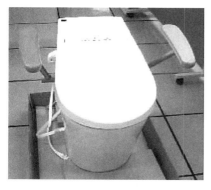

FIGURE 2.5
The new brand TRES (left), the side view of TRES.

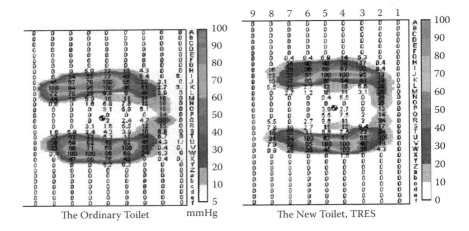

The Ordinary Toilet   mmHg   The New Toilet, TRES

FIGURE 2.6
Body pressure pattern for the ordinary toilet (left) and for the new toilet (right).

implemented an ecosystem concept. We created a product innovation where the toilet has no water tank, which enabled us to reduce water use as much as one-fifth compared with the ordinary toilet. We verified the sitting comfort from the viewpoint of the ergonomic three-dimensional surface and body pressure measurement using FSA (force sensitive application) as well. We obtained a good ergonomic result, as shown in Figure 2.5. You can see that the right picture looks like a very comfortable toilet, because the three-dimensional curved surface fits the human hip shape. In reality, most people felt very comfortable. Figure 2.6 shows the ergonomics of the body pressure pattern. The figure on the left shows the body pressure pattern on the ordinary toilet, which means that the strongest body pressure area is very wide and longer. This implies that the wider part of legs contacted the surface producing an uncomfortable feeling. The right figure shows a narrower contact area for the new toilet, which evokes the comfortable contact without the thigh pressure. The new Kansei toilet was named TRES.

Since the TRES surface was tilted forward 3 degrees and uses two arm rests, standing up became very easy for elderly people. We performed an EMG (electromyograph) of two legs when standing up, and the result showed a reduction of strength needed for standing to one-tenth compared with an ordinary toilet. This case is a typical example of Kansei ergonomics development. The combination of Kansei/affective engineering with ergonomics of easy operation has created a very comfortable toilet with a mechanical innovation (Nagamachi 2008). We realized sitting comfort and ease in standing up, and TRES became very popular among women, because most of them are very aware of environmental problems.

2.3.2 A Mattress That Prevents Bedsores

Every country is becoming an older society, and a plenty of elderly people spend their time lying on a bed, which causes bedsore (pressure sores, decubitus ulcers). Most hospitals in elderly countries like Japan have a big problem with bedsores. The causes of bedsore are as follows:

1. Body pressure on the mattress breaks the skin and disturbs the smooth flow of blood due to distorted veins.
2. The stoppage of blood flow causes skin necrosis.
3. Moisture and bad nutrition facilitate the occurrence of bedsore.

Since numbers 1 and 2 are the most important and risky factors, we attempted to develop a bedsore-preventing mattress from the viewpoint of Kansei/ affective engineering and ergonomics as well. We collected 13 different mattresses that are very popular and used in hospitals. We also asked the product manufacturer Toyobo to create Breathair, a mattress material, of many different densities and heights as well. Breathair is made of polyester and, due to the structure of its three-dimensional entangled polyester "pipes," it has a high rebounding property, as shown in Figure 2.7. We searched by conducting FSA measurements to find the mattress with the lowest body pressure, including Breathair, and selected eight Breathair samples with a sandwich structure. For the research, we employed subjects weighing between 40 kg and 110 kg. Consequently, we evaluated 12 market-ready mattresses made of polyurethane and 8 new samples made of polyester on the 5-point SD scale of Kansei words.

The Kansei scale consisted of the five Kansei words *comfort, don't sink, turn over, good sleep,* and *pleasure,* and the Kansei word scores plus max pressure and mean pressure were analyzed by PCA. Figure 2.8 illustrates the PCA result in which a horizontal axis means Principal Component 1 and a vertical axis

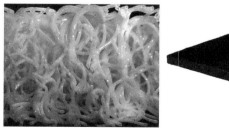 

FIGURE 2.7
(Left) Breathair, made of polyester (a highly rebounding material). (Right) Luckmatair made of Breathair as a bedsore-preventing mattress.

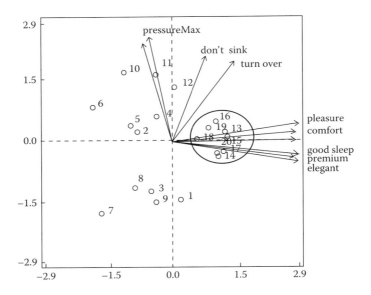

FIGURE 2.8
PCA positioning of 12 mattresses marketed and 8 new candidates.

Principal Component 2. You can see that the Kansei group of *comfortable, good sleep, elegant, premium,* and *pleasurable* gathered around Principal Component 1, which implies the Kansei factor. On the other hand, Principal Component 2 means *pressure max and means, don't sink,* and *turn over.* The circle on Principal Component 1 included all eight new samples with different types of mattress covers and positioned on the middle area for Principal Component 2. The 12 mattresses collected from the market were distributed all round on Principal Component 2, which means that these mattresses made of polyurethane have diversity properties in pressure, namely from light to heavy.

We selected a few mattresses related to lighter body pressure (28 mmHg) and compared these with all mattresses from the market from the viewpoint of a blood flow test with a heavy human body. Figure 2.9 shows the test results related to the blood flow. The upper graph in Figure 2.9 shows the blood flow when on a mattress made of polyurethane, which is most popular for prevention of bedsores in Japan, and the lower illustrates the best blood flow for preventing bedsores with the new product made of Breathair (polyester). The upper graph shows a little blood flow, but the right graph shows high blood flow each time the subject turned over on the mattress. A mattress made of polyurethane surrounds the human body due to its softness, and its softness disturbs the blood flow, even though it has a very low body pressure (Nagamachi et al. 2009).

To verify the effectiveness of the new mattress, named Luckmatair, we donated 10 mattresses to each of five national hospitals and recorded clients' recovery process from bedsores. As the style of medical care was a

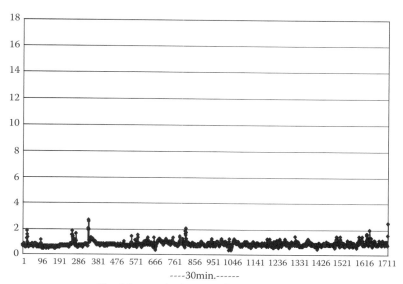

Blood flow on the famous polyurethane mattress

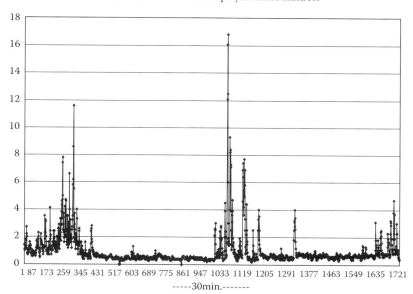

New Kansei Mattress (Vertical index:1ml/min)

FIGURE 2.9
Blood flow process of a polyurethane mattress (upper) and the new mattress (lower).

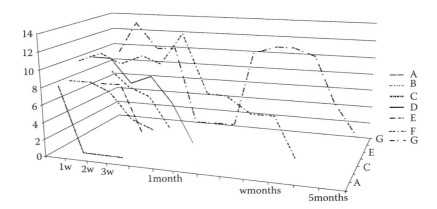

FIGURE 2.10
The recovery process using a new mattress. Five clients recovered soon, but another group with severe bedsore took 1–5 months. (Vertical axis means bedsore score, and horizontal axis means months.)

little different in each hospital, we describe here the results of the new mattresses at Akita Labor Care Hospital. We began observing 15 clients with bedsores in August 2008; however, during observation some patients died. Consequently, we observed 7 clients through the process until complete recovery. The result is illustrated in Figure 2.10. The faster-recovering group recovered completely in a few weeks, but the ones who took longer have taken up to 5 months to reach complete recovery.

Results revealed that the duration for recovery is dependent on the stage of medical severity (bedsore score) ranging from stage I to stage IV. The clients in stages I to III needed just a couple of weeks for recovery. On the other hand, the clients in stages III to IV took a longer time. We started this study in August 2008, and there were no clients with bedsores recorded in the five national hospitals. The bedsores disappeared completely from clients with the use of the new mattress.

The incorporation of ergonomics in Kansei/affective engineering research is concerned with new product property and product quality. The idea, concept, or principle of ergonomics should be implemented in the product development as long as humans have deep concern with the expected product effectiveness.

2.3.3 Soft Computing of Kansei/Affective Engineering

Soft computing is defined as developing a solution by means of information technology, such as artificial intelligence, neural network models, fuzzy logic, and genetic algorithm models. We have developed Kansei computerized software, and some of these are outlined in the following paragraphs.

1. Kansei/Affective Engineering System—Kansei Artificial Intelligence. Kansei artificial intelligence has two objectives. The first is to assist customers in selecting a product that a computerized system suggests will fit the customer's emotion. The second is to assist a designer in making a new design.

   Our first software product is FAIMS (Fashion Design Modeling System), which is concerned with girls' fashion design, and anybody can produce a new fashion design very easily by using the system. FAIMS consists of a word database, a reference engine with a rules base, a design database, design formation database, and a control system. It was developed in 1988 (Nagamachi et al. 1988). In the system, when customers want to select a new product that fits their specific emotion, they input their emotion in keyword(s) into the computerized system. The system processes through the intelligent system and transfers a candidate sample into the display. The system can easily change an image to fit any kind of emotion. When the designers want to use the system, they input their emotional keyword(s) into the system to get the computerized conclusion. If a designer has another idea for a design that fits the word, he or she can manage the relation of the design database with the emotional keywords.

2. HULIS(Human Living System) and HousMall. HULIS is an application of FAIMS into a house design as a shell of the system. It also consists of several databases of artificial intelligence and house design database. It includes all house parts, such as the entrance, Japanese-style room, children's room, Western-style room, kitchen, and bathroom. A client can input emotional keyword(s) related to the house or lifestyle, and the system displays the house design according to the calculation. It is very useful to visualize the design of the customer's house after this process (Nagamachi 1989, Nagamachi and Lokman 2010). A combination of this system with virtual reality technology related to house design has been developed, named HousMall, which enables the visualization of all house designs in three dimensions (Nagamachi 1998).

3. ViVA System. The ViVA System is a combination of Kansei/affective engineering with virtual reality technology applied to a kitchen design. This research is a joint effort of Hiroshima University and Panasonic Electric Works. Nagamachi investigated kitchen layout and design from 10,000 clients, and these data were built into a system. A client will sit down in front of a computer and input the family data and answer the system's inquiries. The answer will then be analyzed by an artificial intelligence system with regard to the family's lifestyle. Finally, the client will input her desires and expectations related to the future kitchen design. The system will calculate

and analyze the data and display the candidate's design related to her imagination. During ViVA operation, all clients were surprised when the new kitchen design appeared on the screen, and they said, "Yes, this is my dream." After this process, all parts of the new kitchen design derived by the system will be delivered to the client's house, and the company renovates the client's kitchen according to the new design proposed by ViVA (Nagamachi et al. 1996).

4. KCOPS (Kansei Cockpit Design System). We developed a new style of exterior design with Komatsu's construction machine in 1983, and the resulting Kansei design has influenced other construction machine makers. At that time we constructed an artificial intelligence system with a cockpit design for construction machinery (KCOPS). We surveyed the drivers' emotional feelings related to the cockpit design, and the surveyed results were incorporated into new design databases of a computerized system for a construction machine. When a driver inputs his Kansei word(s) into the system, the system calculates and displays three-dimensional cockpit graphics on the screen. Guided by the computerized system, the driver can select a cockpit design that fits his emotion.

5. VIDS (Vehicle Interior Design System). The Japanese automobile manufacturer Isuzu asked Nagamachi to build a special artificial intelligence system related to an automobile interior design. A small car's interior is limited, but the manufacturer wanted to make it look wider and broader. The purpose of the system was to give the car designers suggestions on how a driver could feel the interior to be broader and larger than it actually was. We investigated many small car interior designs and measured internal sizes. We conducted a Kansei survey regarding the interior width. After calculation with statistical analysis, the Kansei results were incorporated into an intelligent system from the viewpoint of genetic algorithm. The system is called VIDS. Isuzu has used the system to provide a look of wider space for customers (refer to Chapter 6 and Tsuchiya et al. 1996).

6. WIDIAS (Word Image Diagnosis System). Finally, we describe a very unique artificial intelligence system called WIDIAS, which aims to diagnose whether a brand name would fit the customer's feeling. A manufacturer who creates a new product wants to give a good brand name to the product. However, if the name does not match the customer's Kansei or feeling, it will be unsuccessful in sales. Thus, the name or brand name should fit with most of the customers' emotions. We developed a four-layer neural network model based on the Kohonen Model with fuzzy logic and constructed a hierarchical artificial intelligence system that imitates a human utterance system.

This system is able to judge whether a new brand name gives a good-sounding impression in reference to 40 Kansei scales, and anybody can use this system to find good brand names that fit the customers' emotions (Nagamachi 1995).

References

Nagamachi, M. (1989). *Kansei engineering*, Kaibundo Publishing, Tokyo.

Nagamachi, M. (1995). *Story of Kansei engineering*, Kaibundo Publishing, Tokyo.

Nagamachi, M. (1998). Virtual Kansei engineering applied to house designing, *Human Factors in Organizational Design and Management* VI, 399–404.

Nagamachi, M. (2001). Framework of Kansei engineering and its application to cosmetic product, The 5th International Conference on Engineering Design and Automation, 814–819.

Nagamachi, M. (2008). Perspectives and a new trend of Kansei/affective engineering, *The TQM Journal*, 20 (4), 290–298.

Nagamachi, M., Ishihara, S., Nakamura, M., and Morishima, K. (2009). Kansei engineering and its application to a new mattress preventing bedsore, The 17th World Congress on Ergonomics (2AF0015).

Nagamachi, M. and Lokman, A. M. (2010). *Innovation for Kansei/affective engineering*, CRC Press, New York.

Nagamachi, M., Matsubara, Y., Nomura, J., Sawada, K., and Kurio, T. (1996). Virtual Kansei engineering and an approach to business, *Human Factors in Organizational Design and Management* V, 3–6.

Nagamachi, M., Ito, K., Tsuji, T., Chino, T. (1988). A study of costume design consultation system based on knowledge engineering, *Jap. J. of Ergonomics Soc.*, 24 (5), 281–289.

Osgood, C. E, Suci, G. J., and Tannenbaum, P. H. (1957). *The measurement of meaning*, University of Illinois Press, Champaign.

Tsuchiya, T., Matsubara, Y. and Nagamachi, M. (1996). A development of Kansei engineering system for designing automobile interior space, *Human Factors in Organizational Design and Management* V, 19–22.

3

Psychological Methods of Kansei Engineering

CONTENTS

3.1 Statistical Scaling and Psychological Measurement

Kansei is a psychological phenomenon. In order to utilize Kansei for developing and improving products, it has to be measured and analyzed in psychological terms. Stevens (1946) proposed four classifications of psychological measurement scales.

1. Nominal scale: The object set has only classifications, such as classical, rock, and pop music. Objects classified into the same group have the same label or number. Labels or numbers are given arbitrarily and without order.

2. Ordinal scale: Numbers are given to each object to show the order of the objects. Differences between numbers, or intervals of numbers, are not equal. Typical examples are road races of bicycles or marathons. Often the top group and large main group (peloton) have formed. Between the members of the top group, intervals are small, but the interval between the top group and main group is often large. If there is a top group of three members, intervals between

first, second, and third are small, but the interval between third and fourth is large because the fourth belongs with another group.

3. Interval scale: Objects are ordered and also have constant intervals. Then, differences can be calculated. Interval scale does not have natural zero; in other words, an absence of the object. Examples are temperature, such as Celsius and Fahrenheit, in which $0°$ is arbitrary and does not mean the absence of temperature.

4. Ratio scale: Objects are ordered, have constant intervals, and have natural zero. Examples are height, weight, and length. Ratio can be computed; for example, 20 cm is twice as long as 10 cm.

In psychology, there are two major methods of sensation measurement: magnitude estimation and category method. Magnitude estimation examines the ratios between senses of two stimuli. Category method examines the corresponding category of stimulus or difference of stimuli, from a category set. In the next sections, we will note magnitude estimation and the Likert scale. The Likert scale is one of the category methods and is most often used in various fields of psychology and Kansei/affective engineering.

3.2 Magnitude Estimation

We often express relationships as a comparison of two or more objects, such as a *bigger fish*, 1.5 times as large as the previous one. Magnitude estimation is the measurement method of the ratio between senses by comparisons. In general, a standard stimulus is presented and assigned a numerical value that directly represents its intensity. The number is called a *modulus*. Subsequent stimuli are evaluated by estimated values of perceived intensity ratios in comparison to the standard stimulus. The modulus is assigned as a round number, such as 1 or 100. The comparing stimulus is evaluated in a magnitude ratio, such as 0.7 or 110. Usually, this magnitude is used directly as a measurement value. The representative value is computed between different subjects' ratings. To choose the representative value (i.e., mean or median), the distribution profile of evaluation value should be investigated. If the distribution is not consistent with the normal distribution, then median is more proper.

In subjective estimation of weights, there is a linear relation between $\log_{10}(x)$ of estimation averaged between subjects and $\log_{10}(y)$ of weight. This relation is called Stevens' power law, after Stanley Smith Stevens' paper on various measurements of sense and physical quantity (Stevens 1957).

When applying magnitude estimation to Kansei evaluation, we will propose questions like, "How much more attractive is [stimulus A] than [standard stimulus]?" In some cases the differences between stimuli are too small. When the difference is smaller than a human's *just noticeable difference* (JND), the estimations will fall into noise, or they are meaningless. In such cases, repetition of evaluation is useless.

3.3 Likert Scale

American organizational psychologist Rensis Likert created the Likert scale in his 1920s works on assessment of attitude. For example, asking the question "Satisfied with own work?" on a five-level scale:

1. Strongly agree
2. Agree
3. Neither agree nor disagree
4. Disagree
5. Strongly disagree

Aligning of five levels on the line makes this measure.

Strongly agree / Agree / Neither / Disagree / Strongly disagree
|_____|_____|_____|_____|

With this scale, the evaluation value can be treated as a numerical value at least on the ordinal scale. The answering time is much shorter than on a free-description questionnaire. These advantages are substantial for Kansei/affective engineering. The semantic differential method is a multidimensional evaluation method utilizing the Likert scale on numerous questions.

3.4 Semantic Differential

3.4.1 Theoretical Background of Semantic Differentials

Charles E. Osgood, who was a social political psychologist, developed semantic differentials in the mid-1950s. This is the measurement method for connotative meaning of objects (Osgood, Suci, and Tannenbaum 1957). The concept of Osgood's *connotative meaning* is close to the *signified* (*signifié*) proposed by Saussure, which represents a mental image or an idea of a thing rather than the thing itself.

Osgood created a *representation–mediation process* model of stimuli and human responses, which is a process model of world and connotative meaning. A human receives stimuli from the outer world through his sensory organs. Then stimuli are subjected to the projection process. Next, projections are sent to the integration process. Abstracted representation is sent to the representational mediation process, and the meaning is recognized. The meaning is gradually reduced to concrete responses by reverse-order processes. At last, behaviors are shaped. Although this model was created in precognitivism in psychology, it is still compatible with today's frameworks of cognitive science. Osgood thought if connotative meanings were gradually shaped, they could be resolved into many simpler concepts.

Osgood had to develop a measurement method of connotative meanings. As a social political psychologist (e.g., Osgood and Suci 1955), his interest was in measuring vague and various meanings of words like *Russians, patriots,* and *America.* The semantic differential analyzes connotative meaning by plentiful evaluation words. An adjective is paired with its antonym, like *realistic—idealistic.* Between this pair, the Likert scale was placed (e.g., *realistic* [ ] [ ] [ ] [ ] [ ] [ ] [ ] *idealistic*); the subjects rated their judgment or evaluation on many other pairs of word. Osgood used 50 to 80 word pairs. Usually, a 5-grade or 7-grade scale was used for either *disagree* or *agree.* Since connotative meanings are resolved with numerous words, he named this method *semantic differential.*

Theoretical ideas of semantic differentials are consolidated as these three topics (Osgood and Suci 1955):

1. A continuum is defined by an evaluation word pair that has opposite meanings. Description, judgment, and evaluation processes could be positioned on the continuum. For example, a decision (e.g., XX can be reliable or not) could be quantified on an evaluation word-pair scale like *reliable* [ ][ ][ ][√][ ] *unreliable.*

2. The continuum is equivalent to judgment and evaluation. "XX is not reliable" is an equivalent to the evaluation on the SD scale shown in Figure 3.1. Different continuums that have similar responses can be consolidated. The *reliable—unreliable* pair is highly correlated with pairs like *honest—deceitful, equal—unequal, kind—unkind.* These can be combined to a general factor.

3. From the correlation structure, *semantic space* can be defined. The semantic space is presented in a series of Osgood's studies that have used factor analysis. Some word pairs are strongly correlated, and others are negatively correlated. In addition, some have not been

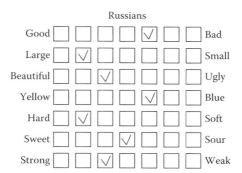

FIGURE 3.1
An example of Osgood's original SD questionnaire on Russians.

correlated. Thus, factorial structures could be extracted with PCA or factor analysis. In mathematical procedures, the semantic space gives features of a large correlation (or variance–covariance) matrix.

3.4.2 Modified SD for Kansei/Affective Engineering

Osgood used polarized antonyms like *beautiful—ugly*. In Kansei/affective engineering, we strongly recommend using denial words like *beautiful—not beautiful* instead of antonyms for two reasons:

1. In the statistical sense, when measuring on a *beautiful—ugly* scale, statistical frequency distribution is distorted toward the beautiful side. Since no manufacturer sets out to create ugly products, only a few products on the market are rated *ugly*. Therefore, most evaluation distribution is placed on the *beautiful* side and a very small distribution on the *ugly* side. Such skewed distribution prevents applying most statistical analysis techniques. In *beautiful—not beautiful*, the distribution becomes symmetrical and close to the normal (Gaussian) distribution.

2. Some antonyms do have not opposite meanings. For example, what word has the opposite meaning of *fashionable*? Deciding exactly the opposite word is difficult. The *fashionable—not fashionable* pair escapes such a semantic problem.

Extracting semantic structure was the aim of Osgood's research. He used factor analysis for extracting structure and argued that evaluation, potency, and activity (EPA) are the general structures for all meanings (Osgood, Suci, and Tannenbaum 1957). Evaluation contains *good—bad*, *stable—unstable*, *happy—sad*, and *beautiful—ugly*. Potency contains *large—small*, *strong—weak*, and *clear—vague*. Activity contains *dynamic—static* and *exciting—calm*. Soon after his 1957 paper, many psychologists examined his EPA model, and they found EPA was not universal as Osgood argued. In many cases, EPA was not clearly separated or more than four factors were extracted. Osgood withdrew his argument of EPA universality in the early 1960s. It is common sense in the psychology field that structures of meaning differ by research objectives.

In Kansei/affective engineering, adjectives, nouns, technical terms, and jargon are often used for SD evaluation. We call these *Kansei words* for a large set of evaluation words.

3.5 Statistical Considerations of SD Data

In a strict classification, SD measurement data are in an ordinal scale. Practically, SD data are treated as a nominal scale in most studies. Thus, arithmetic means and correlations are calculated and then PCA or factor analysis is performed. Hagiuda and Shigemasu (1996) showed that 5- or 7-point SD scale data can provide a statistical distribution of Kansei evaluation data.

When we apply a statistical test, we have to choose parametric tests or nonparametric tests by whether the data distribute along a normal distribution or not. This consideration is taught in introductory statistics and is widely known. Unfortunately, consideration of data distribution is lesser on multivariate analyses than on statistical tests. Some multivariate analysis techniques, such as discriminant analysis, are theoretically derived on normal distribution. Many other techniques also implicitly assume normality of distribution. For example, PCA starts from a variance–covariance matrix or correlation matrix. As we teach in introductory statistics classes, the Peason correlation coefficient is unsuitable for nonnormal distribution. Considering this, the distribution is recommended also on multivariate Kansei evaluation data.

An example of distribution examination is described next. We first examined the raw data of SD evaluations. Figure 3.2 shows the histogram of milk carton and hair treatment evaluation data. The milk carton data has 28 subjects × 25 cartons × 69 Kansei words = 48,300 evaluations. Its mean was 3.05 and the standard deviation was 1.33. Hair treatment has 14 subjects × 43 hair treatments × 39 Kansei words = 23,457 evaluations. Its mean was 3.04 and the standard deviation was 1.23.

Both distributions had their single peak at 3, and their means were also near 3. Since evaluation values were discrete, we can't conclude that these are normal distributions. However, distributions clearly centered on 3 and were symmetrical; means were proper for representative value. For example, when the distribution differs from the normal distribution, median or mode should be used as the representative value.

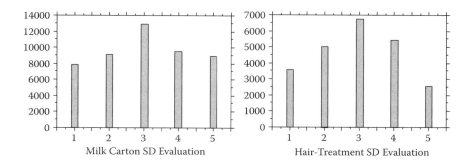

FIGURE 3.2
Distributions of milk carton (L) and hair treatment evaluation data (R).

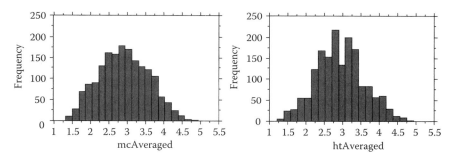

FIGURE 3.3
Distributions of evaluation values (averaged between subjects): milk carton (L) and hair treatment (R).

We considered the average between subjects as the evaluation value. Figure 3.3 shows the average value between subjects of milk carton and hair treatment evaluation.

Concerning milk cartons, the mean was 2.907 and standard deviation was 0.662 on 1725 evaluations (25 cartons × 69 Kansei words). It was skewed at 0.075. Perfect symmetry was 0.0; thus, milk carton data are very symmetrical. Kurtosis (Ku–3), the measure of the peakedness, was –0.562. Kurtosis of normal distribution was 0; thus, data were slightly flatter than the normal distribution. Related to hair treatment, the mean was 2.928 and standard deviation was 0.653. It was skewed at 0.136 and was rather symmetrical. Kurtosis was –0.278, slightly flatter than the normal distribution. We examined the differences from the normal distribution with the Kolmogorov–Smirnov test. There were no significant differences, and the differences presented no difficulty for further analysis.

The above-mentioned analyses were done on flattened data, which means that evaluations on all samples × all Kansei word evaluations were flattened to one-dimensional data. Then we examined p-dimensional multivariate data of p Kansei words. The data are a matrix of the number of samples × p. We examined Kansei evaluations of color samples, ladies' wristwatches, milk cartons, beer cans, and hair treatments. These contained 80 to 100 pairs of Kansei word evaluations. The data were examined with the Q1 and Q2 test of Small (1980) and multivariate skewness and kurtosis tests of Srivastava (1984). These tests show no significant deviations from the multivariate normal distribution.

From these results, we concluded that there are no problems, implicitly or explicitly, in analyzing Kansei evaluation data using multivariate analyses, depending on normality of the distribution.

References

Hagiuda, N., and Shigemasu, K. (1996). Jyunjo tsuki Categorical Data heno inshibunseki no tekiyouni kansuru ikutuka no chuiten, *Shinrigaku Kenkyu*, **67**(1): 1–8. (in Japanese)

Osgood, C. E., Suci, G. J. (1955). Factor analysis of meaning, *Journal of Experimental Psychology*, 50, 325–338.

Osgood, C. E., Suci, G. J., and Tannenbaum, P. H. (1957). *The measurement of meaning*, University of Illinois Press: Urbana.

Small, N. J. H. (1980). Marginal skewness and kurtosis in testing multivariate normality, *Applied Statistics*, **29**(1): 85–87.

Srivastava, M. S. (1984). A measure of skewness and kurtosis and a graphical method for assessing multivariate normality, *Statistics and Probability Letters*, **2**: 263–267.

Stevens, S. S. (1946). On the theory of scales of measurement, *Science* **103** (2684): 677–680.

Stevens, S. S. (1957). On the psychophysical law, *Psychological Review*, **64**(3): 153–181.

4

Psychophysiological Methods

Keiko Ishihara

CONTENTS

4.1 Product Development

Since 2000, we have been involved in the development of many products, and we have recognized that Kansei/affective engineering and ergonomics are inseparable. Attractive products cannot be made with only ergonomic considerations, and Kansei/affective engineering provides eloquent answers. Thus, we are proclaiming the need for Kansei/affective ergonomics.

In this chapter, we introduce methodologies of Kansei/affective engineering with the explanation of two examples of Kansei/affective ergonomics. One is the development of a pen-grip shaver, and the other is an evaluation of washer-dryer machines.

4.1.1 Pen-Grip Shaver

Home electrical appliances are changing from low-priced, mass-produced items to higher-priced, higher-function items. The mechanism of a typical electric shaver involves an inner blade, which moves inside of the mesh outer blade. Thus, when more pressure is added to the face, the shaving becomes poorer and load is added to the inner blade. The conventional stick-shaped shaver tends to force users to apply more pressure to their faces. Sanyo engineers thought that bending the shaver head and providing it with a pen grip, like a T-shaped razor, would address this problem. We have verified the idea with experiments and measurements.

4.1.2 Washer-Dryer

Recently, washer-dryer machines with horizontal or slanted drums have become popular in Japan. Traditionally, Japanese washing machines have had vertical drums, and these types are still common. Users of vertical-drum washers have to bend their backs and stretch their arms to put in and take out laundry. Meanwhile, in Europe, horizontal-drum washing machines have long been popular. This type requires a crouching posture for putting in and taking out laundry because of its lower height.

The new washer-dryers, with horizontal or slanted rotational axes of the drum, have rather different mechanisms from vertical-drum washing machines, and thus require a new and different mechanical design. Thus, the shape of the washing machine was greatly changed; to make loading operations easier, the door position was modified.

In this research, physical loads and usability of the new washer-dryer machine, the traditional Japanese drum, and European washing machines were compared. This comparison was performed using subjective evaluations, 3-D motion capture, and estimations of body part loads using a human kinetics computer model.

4.2 EMG Measurement of Pen-Grip Shaver

We examined two types of shavers: a conventional stick and a prototype of a new pen-grip shaver. These two shavers had the same grip part, and their grip length and diameter were identical. The stick shaver had its head at 15° from the grip, and the pen grip prototype had its head at 80° (Figure 4.1).

1. Electromyogram (EMG) measurements. The experimental problem was the difference in EMG between the NS1 (existing stick shaver) and the pen-grip prototype (reformed NS1). Electrodes of EMG were attached on the *flexor digitorum superficialis* and on the *flexor digitorum profundus*, with bipolar derivation. Measurements were conducted with two channels. The measurement device was a Biopac MP30 (Biopac Inc.) with a sampling rate of 500Hz.

2. Pressure-to-face measurements. A piezo pressure sensor was attached behind the blade of the shaver. The factor in this experiment

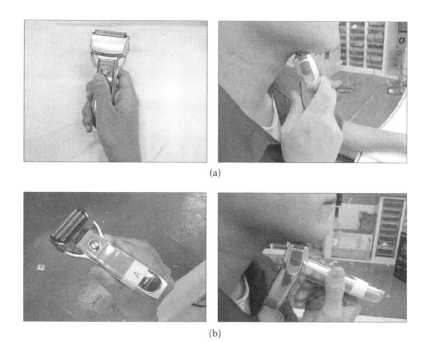

FIGURE 4.1
(a) Stick shaver and (b) pen-grip prototype.

was the difference in pressure to the human face between the stick shaver and the pen-grip prototype. Measurements were also made with the Biopac MP30.

3. Instruction to the participants. Instructions were given to the participants, including application to the face and shaving direction. The task was to move the shaver three times at seven different sites: the middle under the chin, right and left of it, on the chin, under the nose, right cheek, and left cheek. The subjects were seven men in their 20s.

4. Consequence of EMG measurement. The pen-grip prototype had smaller voltages (see Figure 4.2). The upper panel shows stick shaver

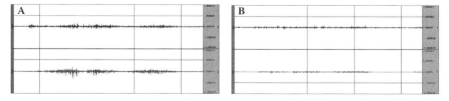

FIGURE 4.2
EMG examples in the middle, under the chin: (A) stick shaver and (B) pen-grip shaver.

data and the lower panel shows pen-grip prototype data for the same subject during shaving the middle under the chin. The upper row corresponds to the *flexor digitorum superficialis*; the lower row corresponds to the *flexor digitorum profundus*. One tick on the *y*-axis is 1mV; one tick on the *x*-axis is 2s.

EMG integral values (mV × sec/500 (Hz)) of the two shavers (sum of the seven sites) were compared using measurements from the seven participants. The ratios between the pen-grip prototype and stick one (average between subjects) were 0.60 at the *flexor digitorum superficialis*, 0.95 at the *flexor digitorum profundus*, and 0.78 combining both muscles. Thus, a 22% EMG reduction was observed in the pen-grip prototype.

The statistical distribution of differences between pen grip and the stick was not a standard distribution, according to the Shapiro-Wilk W test. These were paired data, because the same subject used both shavers. Thus, we used the Wilcoxon signed-rank test, a nonparametric test of paired data. Using this test, the difference in EMG integral value between two shavers was statistically significant ($p < 0.0001$).

5. Consequence of pressure-to-face measurement. As shown in Figure 4.3, the pen-grip prototype had a lower pressure. The left panel shows stick shaver data and the right panel pen-grip prototype data for the same subject. One tick on the *y*-axis is 50 mV. The pressure integral values (mV × sec/500(Hz)) of the two shavers (sum of seven sites) were compared in measurements from the seven participants. The ratio between the pen-grip prototype and the stick (average between subjects) was 0.15. Thus, an 85% pressure reduction was demonstrated using the pen-grip prototype. The statistical distribution of differences between the pen-grip and stick shavers was a standard distribution. Thus, we used the paired t-test; the difference in pressure integral value between two shavers was statistically significant ($p < 0.0001$).

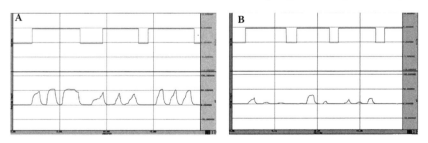

FIGURE 4.3
Examples of pressure to the face; middle, right, and left under the chin: (A) stick shaver and (B) pen-grip shaver.

FIGURE 4.4
Commercial realization of the pen-grip shaver (SANYO SV-GS1).

From the experiment, the pen-grip prototype reduced 22% of the forearm EMG and provided an 85% reduction in pressure to the face. Statistical tests demonstrated the significance of these reductions. As a result, the pen-grip shaver was developed and launched in March 2008 and has generated large sales volumes at a fairly high price (around JPY9500, or US$100). It was named a "Good Practice of Ergonomics" by the Japanese Ergonomics Society. In the spring of 2009, two derived variations were added to the lineup.

4.3 Musculoskeletal Model and Kansei Engineering of Washer-Dryer

4.3.1 Washer Evaluation Experiment Method

In this experiment, we requested that participants take laundry out of the machines. As a model laundry load, two towels were placed at the bottom of the drum, and two blankets, each 1.6 kg, were placed on the towels. These items were dry. The participants were asked to open the door, take out the laundry piece by piece, put the items into a basket that was placed on the floor, and then close the door. The participants were 12 females, ages 20 to 43. Four subjects were short (148–153 cm); five subjects were around the Japanese female average of 158 cm; and three subjects were taller, around 165 cm.

FIGURE 4.5
Laundry machines: (A) European washer AWD-500, (B) vertical-drum washer ASW-800, and
(C) slanted-drum washer-dryer AQ-1.

Three laundry machines were used, as shown in Figure 4.5: a European
box-shaped washing machine (Sanyo AWD-500, referred to below as the EU
type), a typical Japanese vertical-drum washing machine (Sanyo ASW-800,
referred to as vertical drum), and a slanted-drum, fully automatic washer-
dryer machine (Sanyo AQ-1; referred to as slanted drum). The height to
the center of the opening was 47.5 cm for the EU-type machine, 90 cm for the
vertical-drum machine, and 81 cm for the slanted-drum machine. Note that
the opening of the vertical-drum machine faced straight up, meaning that
laundry had to be lifted higher than the actual height of the door.

4.3.2 Washer Subjective Evaluation

A subjective evaluation was carried out by asking the participants a set of
questions each time their required task was completed. Of the questions, five
were related to fatigue, five to usability, and a final question to the general
usability of the washing machine. Table 4.1 lists the questions asked. Each
question was answered on a 5-point scale. One-way analysis of variance and
post-hoc test (Tukey-Kramer HSD) were used.

The results indicated that the slanted-drum machine has the highest evaluations for all questions. On fatigue and usability questions, we found that
the slanted-drum machine and vertical-drum machine scored significantly
better than the EU-type machine. On taking out laundry from the bottom,
the slanted-drum was significantly better than the vertical drum.

TABLE 4.1

Questions for Subjective Evaluation

1. How tired does your entire body feel?
2. How tired are your neck and shoulders?
3. How tired are your upper arms?
4. How tired is your back?
5. How tired are your knees?
6. How easy was it to push the Door Open button?
7. How easy was opening and closing the door?
8. How easy was it to check inside the drum?
9. How easy was it to insert your hand or arm inside the drum?
10. How easy was it to take out laundry?
11. How easy was the machine to use?

4.3.3 Working Posture Measurements by Motion Capture and Analysis of Joint Angles

We measured working postures with the Proreflex 3-D motion-capture system (Qualisys Inc., Sweden), which has five infrared cameras. Using this motion-capture system, we measured working postures in terms of coordinate values for various parts of the body. The sampling rate was set at 120 samples and the spatial resolution setting during measurements was 5–10 mm. Figure 4.6 shows the posture of a subject with a height of 158 cm (the average for Japanese women) during maximum bending of the body when removing a towel from the drum.

Markers were set at 15 locations on the subject's body: head, left and right shoulders, left and right elbows, back (dorsal) of each hand, left and right greater trochanter, left and right knees, left and right ankles, and left and right toes (on the subject's slippers).

Using data from the motion capture, we measured and analyzed the angle formed by the knee, greater trochanter, and shoulder. This angle was 100° (average between subjects) for the slanted drum, 114° for the vertical drum, and 64° for the EU type (Figure 4.6). Because standing posture is close to 180°, the larger angle was the better. One-way analysis of variance indicated that differences between machines were significant ($F(2,33) = 37.622$, $p < 0.0001$). The results of the HSD test revealed a significant difference between the slanted-drum and EU-type machines and between the vertical drum and EU-type machines ($p < 0.05$).

The angle formed for the slanted drum was $110/64 = 1.71$ times larger than that of the EU type, which can be interpreted as a 70% improvement. For the EU type, the capture screen showed that laundry could not be put in or taken out without squatting completely. This is likely the reason for the poor

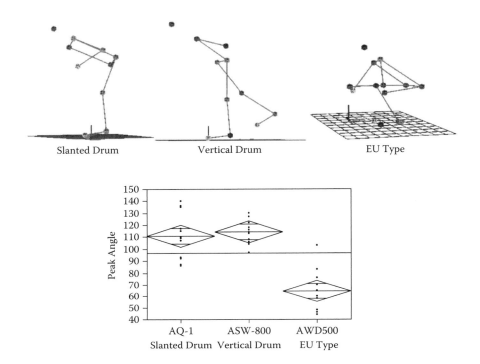

Slanted Drum   Vertical Drum   EU Type

FIGURE 4.6
Posture during maximum bending of body (158-cm young female) and graph of angles formed by the knee, greater-trochanter, and shoulder for different machines.

evaluations given to the EU-type washing machine for the questions "How tired does your entire body feel?" "How tired are your knees?" and "How easy was the machine to use?" The vertical drum provided a posture closer to the vertical stance than that of the slanted drum, but since the vertical drum was deep, almost all of the participants had to reach the towel at the bottom of the drum by raising a foot off the ground and stretching inside the drum. This is why the vertical drum was poorly evaluated with respect to the question "How easy was it to take out laundry?" The relationship between the subjective evaluation and working posture was therefore clarified by measuring body posture through motion capture and calculating the angle of body bending.

We have shown that the vertical drum required an off-balance posture. The entire body load at this time cannot be estimated solely on the basis of coordinates and angle data obtained through motion capture. The load on the lumbar vertebrae that cannot be directly measured is also an important factor. Accordingly, giving due consideration to the mass of various parts of the body, we attempted to estimate such loads using a kinematic model.

4.3.4 Static Load Estimates Using a Kinematic Model

We estimated the load on various parts of the body using a kinematic model. To perform these calculations, we used the 3-D Static Strength Prediction Program (3-D SSPP) developed by a team led by Professor Don Chaffin at the University of Michigan. Chaffin has been researching kinematic models of the human body and applying them to posture analysis of production lines for about 30 years. As shown in Figure 4.7, the Chaffin model features a human body with a basic structure consisting of seven links. These links are the forearm, upper arm, torso (shoulder to lumbar vertebrae), sacral vertebrae to pelvis, femoral head to knee, shank, and foot.

The model takes the following values as major parameters: load, own weight, height, and joint coordinates. The center of gravity is determined by each part's size and weight. As an example in [1], a load of 5 kg (49 N) is held in the hand, with the combined weight of the forearm and hand being 15.8 N. The upper arm, from the elbow up, holds this load with force R_{elbow} in a stationary position. This can be expressed as -49 N $- 15.8$ N $+ R_{elbow} = 0$, then R_{elbow} is 64.8 N in the upward direction.

The rotation moment M_E is in equilibrium with the (center of gravity of the forearm × the weight of the forearm and hand) + (length from the joint to the grip × the load). This can be expressed as 17.2 cm (-15.8 N) + 35.5 cm (-49 N) + $M_E = 0$. Thus, $M_E = 2011.3$ Ncm (20.113 Nm). This assumes the forearm to be in a horizontal position, so any deviation from the horizontal in the form of $-q_E$ will give a result of $\cos\theta_E(M_E)$. For the upper arm, the upward pulling force at the shoulder can be expressed as RS = $W_{UA} + R_{elbow}$, where W_{UA} is the upper arm's own weight. The torque at the shoulder can be expressed as MS = $-(SCM_{UA}) (W_{UA}) - (S_E) (R_{elbow}) - (M_E)$, where SCM_{UA} is the distance from

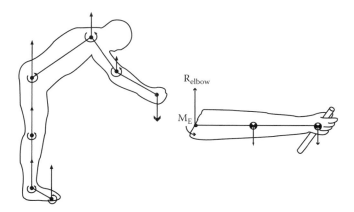

FIGURE 4.7
Body links (entire body) and forearm.

the shoulder to the center of gravity of the upper arm, and S_E is the length of the upper arm. Lowering the upper arm from the horizontal gives a result of cos θ(M_S). In this way, load and joint moments can be progressively calculated for various parts of the body.

Using this model, we estimated the pressure (N) on the disk between the fourth and fifth lumbar vertebrae and the maximum voluntary contraction (%MVC) for the muscles involved in the elbow, hip, knee, and ankle joints for the posture corresponding to maximum bending of the body (for a 158-cm, 53-kg participant). The participant's height and weight were used for the estimation. Referring to Table 4.2 and Figure 4.8, the slanted drum exhibited smaller muscle strengths, except for the hips. For the vertical drum, the pressure on the intervertebral disk was smaller than that of the other two machines, because the back was not bent so much. On the other hand, laundry cannot readily be removed from the bottom of a vertical drum without raising one foot, so that the load on the ankle of the other foot exceeded 100%. The load on the hip and knee was likewise high.

Summing individual %MVCs and comparing the overall %MVC between the different machines revealed that the slanted drum had the smallest value with a muscle load about 60% less than that of the vertical drum. Comparing the slanted drum and the EU type, the latter exhibited a smaller load on the hip but 2.36 times the load on the knee, because a squatting posture must be taken. The above results demonstrate that the slanted drum provided improved posture. We have shown practical examples of the improvement of commercial products with Kansei ergonomics. Consumers may demand

TABLE 4.2

Values Estimated by the Model (158-cm young female)

Subject: 158 cm/53 kg	L4/L5 Comp	Elbow	Hip	Knee	Ankle	Sum (%MVC)	Sum (%MVC)/400
Slanted drum	1732	12	54	25	25	116	0.29
EU type	1801	17	31	59	26	133	0.3325
Vertical drum	1431	8	75	91	110	284	0.71

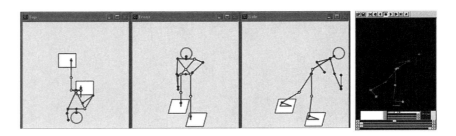

FIGURE 4.8
Calculation screen for the vertical drum (158-cm young female).

both scientific evidence and attractiveness in products. We believe Kansei/ affective ergonomics is an indispensable methodology for successful product development and improvement.

4.4 Comfort and EEG

In Kansei/affective engineering, EEG (electroencephalogram) measurement and analysis technique were pioneered by the late Professor Tomoyuki Yoshida. It is well known that the alpha wave element of EEG increases during quiet rest. Yoshida measured frequency and amount of fluctuations of alpha wave (8 to 13 Hz) from the frontal lobe (Fp1, Fp2) (fluctuation is called *yuragi* in Japanese). He unveiled the relationship between them and tranquil comfort (or relaxed feeling). The outline of his analysis method is as follows. At first, alpha wave elements were extracted by digital frequency filtering. Then, the frequency of each cycle was measured with zero-crossing. Frequencies of the samples in a period shape a statistical distribution. This distribution shows the fluctuation of alpha wave frequency, which has its mode (most frequently measured value) as its peak. The power of the fluctuation distribution decreases along with the lower and higher frequency from the mode. With the Log-Log plot, Log_{10} (power) scale as y-axis and Log_{10} (frequency) as x-axis, the distribution shape becomes the conjunction of two lines. The slope of a lower frequency distribution is less steep until the point between -1.3 (= Log_{10} (0.05 Hz)) and 0 (= Log_{10} (1.0 Hz)). A higher frequency part has steeper slope.

From many experiment results with more than 600 participants, Yoshida found that a gradient of the plot on a lower frequency domain is strongly associated with comfort and relaxed feeling. A steeper gradient, approaching

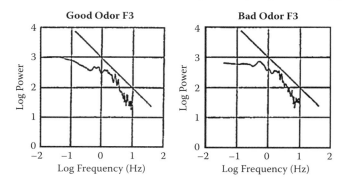

FIGURE 4.9
Fluctuation distribution of alpha wave on smell experiment (Yoshida, T. (1990). *Journal of the Acoustical Society of Japan,* **46**(11), 914–919).

the straight line with a higher frequency domain, is strongly associated with subjective comfort feeling. Less comfort is associated with a flat plot. Smell, visual image, low-frequency noise, facial massage, and TV viewing are among some of his experiments. From his many research results, he constructed a basic emotion model composed of two axes, arousal feeling (excitement/calm) and comfort/discomfort (Yoshida 2000).

In 2000, Yoshida found alpha wave fluctuation associations not only with comfort, but also with enjoyment, which is comfort of excitement. His discoveries are based on enormous measurements and statistical considerations of broader frequency range of fluctuation, while many other attempts had been focused on narrower frequency zone (i.e., 1/f). We mourned Yoshida's unforeseen death, but his theory should be further developed and utilized.

References

Chaffin, D.B., Andersson, G.B.J., and Martin, B.J. (2006). *Occupational Biomechanics* (4th edition), Wiley.

Ishihara, S., Ishihara, K., Nagamachi, M., Sano, M., Fujiwara, Y., Naito, M., and Ozaki, K. (2008). Developments of home electric appliances with Kansei ergonomics—SANYO cases: Kansei and Kinematic considerations on washer-dryer and electric shaver, *Proc. of the 2nd European Conf. on Affective Design and Kansei Engineering*, Lund University Press, (CD-ROM).

Ishihara, S., Ishihara, K., Nagamachi, M., Sano, M., Fujiwara, Y., and Naito, M. (2009). Kansei ergonomic product development of washer-dryer and electric shaver, *Proc. of World Congress of International Ergonomics Association 2009*, Beijing (CD-ROM).

Yoshida, T. (1990). The measurement of EEG frequency fluctuation and evaluation of comfortableness, *Journal of the Acoustical Society of Japan*, **46**(11), 914–919. (in Japanese)

Yoshida, T., and Iwaki, T. (2000). The study of early emotion processing in the frontal area using a two-dipole source model, *Japanese Psychological Research*, **42**(1), 54–68.

5

Statistical Analysis for Kansei/Affective Engineering

Mitsuo Nagamachi

CONTENTS

Multivariate analyses play principal roles in Kansei/affective engineering. Kansei evaluation data have multidimensional characteristics because the nature of Kansei is multidimensional, as mentioned in Chapter 3. When we utilize multivariate analyses in Kansei/affective engineering, we often take the following steps:

1. Principal component analysis (PCA) is used to obtain Kansei structures. In some cases, multidimensional scaling method and factor analysis are also used.

2. Cluster analysis classifies samples as groups with similarity of their evaluation. From the results, we obtain the sample clusters, each of which has a different decisive design structure. The neural network-based method performs more precise clustering than traditional algorithms.

3. To obtain relationships between Kansei and design details, several analyses will be tested to determine what type of appearances and functionalities are produced for Kansei information.

   In most cases of Kansei/affective engineering, the design elements are expressed as categorical variables. Then, we used quantification theory Type 1 (QT1) for analyzing relationships between Kansei evaluation and design elements. QT1 is a variation of regression analysis that deals with continuous variables. QT1 deals with categories for explanatory variables.

   Often in cases of real product development, we proceed with a huge number of design elements—too many to analyze using general multivariate analyses. We use the partial least squares method in such cases because it can process a larger number of explanatory variables in the model.

4. Local regression method is used to take account of nonlinear relationships between design elements and Kansei evaluation in the statistical model. This is a useful tool for visual investigation of uneven local relations between variations of a design element and Kansei evaluation.

5. We use correspondence analysis or quantification theory Type 3 to map the variations of the design element to visualize the results. The local regression method used with these analyses provides three-dimensional representations of relationships between design and Kansei evaluation.

5.1 Principal Component Analysis

5.1.1 Meaning of the Principal Component

Given a matrix of Kansei evaluation data $\{y_1, y_2, y_i...,y_n\} \times m$ samples for evaluation in Kansei/affective engineering, the value y is usually the average evaluation among the participants. Because y_i contains the evaluation values for m samples, it is referred to as vector y_i and contains the evaluation values of all samples as its elements.

Figure 5.1 shows the general format of the evaluation data. Subscripts are listed in column and row order (e.g., $y_{RowColumn}$) to correspond to data tables used in Kansei/affective engineering, which means the reverse of the usual notation.

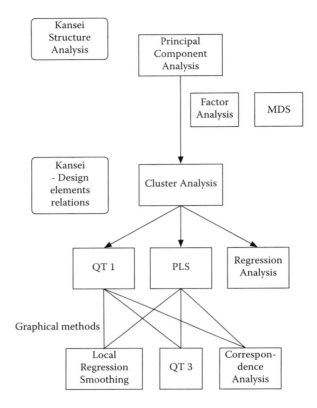

FIGURE 5.1
A scheme of Kansei evaluation data analysis with multivariate analysis techniques.

The aim of PCA in Kansei/affective engineering is to obtain a linear combination of variables that summarizes an n-dimensional distribution (e.g., $n = 80$ for 80 Kansei words), using a lower-dimensional space (Figure 5.2). Excellent computer graphics make it easy to visualize data in three dimensions or fewer. Some applications permit viewing multidimensional data by selecting three variables. In general, one can intuitively understand a maximum of three variables at a time.

Kansei/affective engineering is based on the fundamental view that human Kansei is so complex that new measures are required to analyze it. The measurement of many Kansei words results in multidimensional evaluation data. We describe the characteristics of the evaluation data using fewer dimensions or variables, so that people can understand it. In other words, we compress the dimensionality in order to understand the data structure. This is the reason for using PCA in Kansei/affective engineering.

Figure 5.3 shows an example of Kansei evaluation data with two variables ($n = 2$), where principal components are represented by lines PC_1 and PC_2. In this case, all the original data points can be plotted on the plane because the

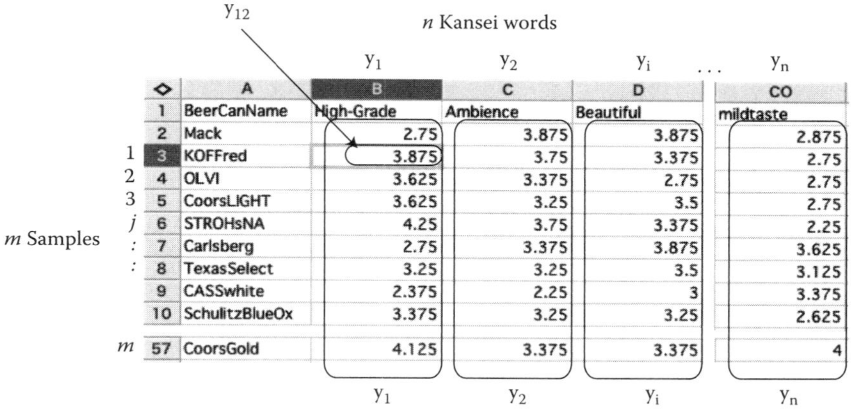

y12
n Kansei words
y1
y2
yi
...
yn
A
B
C
D
CO
1 BeerCanName High-Grade Ambience Beautiful mildtaste
2 Mack 2.75 3.875 3.875 2.875
1 3 KOFFred 3.875 3.75 3.375 2.75
2 4 OLVI 3.625 3.375 2.75 2.75
3 5 CoorsLIGHT 3.625 3.25 3.5 2.75
j 6 STROHsNA 4.25 3.75 3.375 2.25
: 7 Carlsberg 2.75 3.375 3.875 3.625
: 8 TexasSelect 3.25 3.25 3.5 3.125
9 CASSwhite 2.375 2.25 3 3.375
10 SchulitzBlueOx 3.375 3.25 3.25 2.625
m 57 CoorsGold 4.125 3.375 3.375 4
m Samples
y1
y2
yi
yn

FIGURE 5.2
Data format for PCA.

y2
PC2
PC1
y1
Ambience
High Grade
1.5
1.0
0.5
0.0
−0.5
−1.0
−1.5
−1.5
−1.0
−0.5
0.0
0.5
1.0
1.5

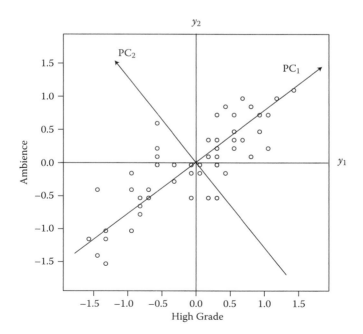

FIGURE 5.3
Geometrical representation of principal components.

data are based on two variables. Each point represents an object sample for evaluation (i.e., each element value (y_{ij}) of the vector y_i is plotted as a point). Actually, the average of y_i, $\bar{y}_i$ is subtracted from each value; the points represent $y_{ij} - \bar{y}_i$.

The points for m samples are distributed from the lower left to the upper right. Line PC_1 drawn diagonally through the center of the distribution as a new axis in Figure 5.2 represents the summarized characteristics of the data distribution with minimum information loss. Axis PC_1 is a composed variable, referred to as the first principal component. PC_1 is a linear composition of the original two variables: that is, a sum of weighted (multiplied by the angle) original variables. PC_1 is a single new variable that is a substitution for two other variables. The data vary a lot from PC_1. Because information loss along PC_1 is minimal, finding another axis is possible, along which the variance of the data would be the greatest.

If the data set is sufficiently large, the data will be generally distributed in the form of a football and PC_1 will pass through the center of the longer axis. One aim of PCA is to find several axes that reduce more than two variables into one, with a minimum loss of information, and to understand the characteristics of the axes.

5.1.2 Search for Maximum Variance

We determine a line PC_1 so that it lies along the maximum variance of the original data. There is a mathematical procedure for determining such an inclination.

We can find such a line by maximizing

$$\sum_{j=1}^{n} (OZ_j)^2$$

as shown in Figure 5.3. Here, PC_1 is the new axis, composed of original variables, and is represented as z. Z_i is a crossing point on a perpendicular line from ith data toward axis z. The task is to maximize the sum of the squared distance OZ_i between Z_i and the origin O (Figure 5.4). In other words, the task is to

$$\text{maximize} \left(\sum_{j=1}^{n} (OZ_j)^2 \right) \tag{5.1}$$

To perform the maximization in Equation (5.1), we first express how Z changes when the original data (dots in Figure 5.3) are projected onto the new axis z. The axis is rotated; the first dimension y_{1j} and the second dimension y_{2j} of the original axis y_i are transformed with rotation parameters l_1 and l_2. The equation below is a weighted sum of composite variables.

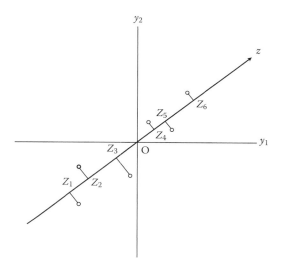

FIGURE 5.4
Maximization of a sum of squared distances.

$$Z_j = l_1 y_{1j} + l_2 y_{2j} \qquad (5.2)$$

There are six data points and two dimensions shown in Figure 5.3. This is the case for the number of evaluation samples $m = 6$ and the number of Kansei words $n = 2$. These data are transformed into $\mathbf{z}$, a vector of Z_j, with the following equation:

$$
\mathbf{z} =
\begin{bmatrix}
z_1 \\
z_2 \\
z_3 \\
z_4 \\
z_5 \\
z_6
\end{bmatrix}
=
\begin{bmatrix}
l_1 y_{11} + l_2 y_{21} \\
l_1 y_{12} + l_2 y_{22} \\
l_1 y_{13} + l_2 y_{23} \\
l_1 y_{14} + l_2 y_{24} \\
l_1 y_{15} + l_2 y_{25} \\
l_1 y_{16} + l_2 y_{26}
\end{bmatrix}
=
\begin{bmatrix}
y_{11} & y_{21} \\
y_{12} & y_{22} \\
y_{13} & y_{23} \\
y_{14} & y_{24} \\
y_{15} & y_{25} \\
y_{16} & y_{26}
\end{bmatrix}
\begin{bmatrix}
l_1 \\
l_2
\end{bmatrix}
= \mathbf{Yl} \qquad (5.3)
$$

Although y_{1j} and y_{2j} included in the matrix $\mathbf{Y}$ contain original data, they are averaged deviation data. Each of them is the result of subtraction of the average from the original value, so that the average of $\mathbf{Y}$ is 0 (see Figure 5.2). The transformation with l_1 and l_2 is briefly described using $\mathbf{Y}$ as

$$\mathbf{Z} = \mathbf{Yl} \qquad (5.4)$$

We performed these calculations using the R statistical software package. We substitute -0.571428 and 0.50375 for y_{11} and y_{21}, and 0.79 and 0.61 for l_1 and l_2, respectively.

```
y11 <- -0.571428
y21 <- 0.59375
0.79 * y11 + 0.61* y21
[1] -0.08924062
```

Calculation with vectors can be performed in R, as shown above. The function cbind composes scalars into columns; that is, it combines them horizontally into a vector. The operation %*% is an inner product of vectors (i.e., a summation of multiplication of elements).

```
y <- cbind(y11,y21)
y
          y11       y21
[1,] -0.571428 0.59375
l <- c(0.79, 0.61)
l
[1] 0.79 0.61
y %*% l

     [,1]
[1,] -0.08924062
```

As mentioned above, the goal is to maximize

$$\sum_{j=1}^{n}(OZ_j)^2$$

or maximize the variance of **z**. The former is a maximization of the sum of the squared distanced from the origin. The latter follows the definition of variance because the average of the data (the averaged deviation of the original data) is zero.

The variance is represented by the following equation, which is a definition of the variance of z_i, the elements of the vector **z**.

$$Var(\mathbf{z}) = \frac{1}{m}\sum_{j=1}^{m}(z_j - \bar{z})^2 \tag{5.5}$$

The problem can be solved by finding a vector **l** that maximizes the equation above, although this requires that $l_1^2 + l_2^2 = 1$. In other words, we add a condition,

$$\mathbf{l}^T\mathbf{l} = 1. \tag{5.6}$$

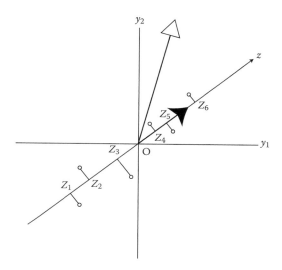

FIGURE 5.5
Orientation of the vector l.

A larger l magnifies **Y** and makes the variance larger with no limit in all directions. For example, even the white arrow in Figure 5.5 may become a candidate orientation, although it is to the left of the orientation of the maximum variance because of magnification. We must fix the length of the vector l to find the correct orientation.

Based on this, the problem becomes that of maximizing Equation (5.5) under the condition of Equation (5.6). Before applying Lagrange's method of undetermined multipliers, we simplify Equation (5.5).

$$Var(\mathbf{z}) = \frac{1}{m}\sum_{j=1}^{m}(z_j - \bar{z})^2$$

$$= \frac{1}{m}\sum_{j=1}^{m}\left\{\left(l_1 y_{1j} + l_2 y_{2j}\right) - \left(l_1 \bar{y}_1 + l_2 \bar{y}_2\right)\right\}^2 \qquad (5.7)$$

$$= \frac{1}{m}\sum_{j=1}^{m}\left\{l_1\left(y_{1j} - \bar{y}_1\right) + l_2\left(y_{2j} - \bar{y}_2\right)\right\}^2$$

$$= l_1^2 Var(y_1) + 2l_1 l_2 Cov(y_1, y_2) + l_2^2 Var(y_2)$$

Here, we represent both Cov (covariance) and Var (variance) as *V*. The reason for this is the covariance is a variance between two different data, and the variance is a covariance between itself; they share the same procedure.

When V_{12} represents the covariance between y_1 and y_2, Equation (5.7) becomes simpler:

$$Var(\mathbf{z}) = l_1^2 V_{11} + 2l_1 l_2 V_{12} + l_2^2 V_{22} \qquad (5.8)$$

Lagrange's method of undetermined multipliers is represented as a function $F = f - \lambda g$. This can be solved by considering f to be the equation to be maximized and setting the condition $g = 0$. Specifically in this case, the equation becomes

$$F = l_1^2 V_{11} + 2l_1 l_2 V_{12} + l_2^2 V_{22} + \lambda \left(1 - l_1^2 - l_2^2\right) \qquad (5.9)$$

5.1.3 Solution of the Lagrangian Function

Equation (5.9) contains three variables, l_1, l_2, and λ. We differentiate the equation partially with respect to each of these three variables. We set each variable to 0 to obtain a set of three simultaneous equations and then solve them. The partial differentials with respect to each variable are

$$\frac{\partial F}{\partial l_1} = 2\left(l_1 V_{11} + l_2 V_{12}\right) - 2\lambda l_1 \qquad (5.10)$$

$$\frac{\partial F}{\partial l_2} = 2\left(l_1 V_{12} + l_2 V_{22}\right) - 2\lambda l_2 \qquad (5.11)$$

and

$$\frac{\partial F}{\partial \lambda} = 1 - l_1^2 - l_2^2 \qquad (5.12)$$

Set each equation to 0 and solve the simultaneous equations that result. The covariances v_{12} and v_{21} are the same. Then, transposed terms simplify, to obtain:

$$V_{11} l_1 + V_{12} l_2 = \lambda l_1, \qquad (5.13)$$

$$V_{21} l_1 + V_{22} l_2 = \lambda l_2 \qquad (5.14)$$

and

$$l_1^2 - l_2^2 = 1 \qquad (5.15)$$

Equation (5.15) is the same as the original condition.

Equations (5.13) and (5.14) can be expressed succinctly by defining a matrix **S**,

$$\mathbf{S} = \begin{bmatrix} V_{11} & V_{12} \\ V_{21} & V_{22} \end{bmatrix}, \quad \mathbf{1} = \begin{bmatrix} l_1 \\ l_2 \end{bmatrix}$$

This results in

$$\mathbf{Sl} = \lambda \mathbf{l} \tag{5.16}$$

The matrix S is a variance of covariance between the variables v_1 and v_2. It is a symmetric matrix because v_{12} is the same as v_{21}, as mentioned above. The task of solving Equation (5.16) becomes that of finding eigenvalue λ and eigenvector **l** of matrix **S**.

This is performed by finding an eigenvalue and an eigenvector of a matrix that is built by variance of covariance between data. (Note that the expansion between Equations (5.6) and (5.16) is based on Mitsuchi 1997, which is an excellent reference for further study.)

5.1.4 Eigenvalue and Eigenvector

Assume a vector **w** on a two-dimensional plane. For further convenience in calculation, let **w** be a column vector (i.e., with its elements aligned vertically). This vector can represent a point on a two-dimensional plane composed by variables x and y.

$$\mathbf{w} = \begin{bmatrix} x \\ y \end{bmatrix} \tag{5.17}$$

Assume a square symmetrical matrix **A**.

$$\mathbf{A} = \begin{bmatrix} 3 & 1 \\ 1 & 3 \end{bmatrix} \tag{5.18}$$

Multiplying **w** by **A** results in

$$\mathbf{Aw} = \begin{bmatrix} 3 & 1 \\ 1 & 3 \end{bmatrix} \begin{bmatrix} x \\ y \end{bmatrix} = \begin{bmatrix} 3x+y \\ x+3y \end{bmatrix} \tag{5.19}$$

Using R, consider how the square matrix **A** affects the space. The following program creates 20×20 points around the point $(0, 0)$. Let the 21×21 matrix be **X**.

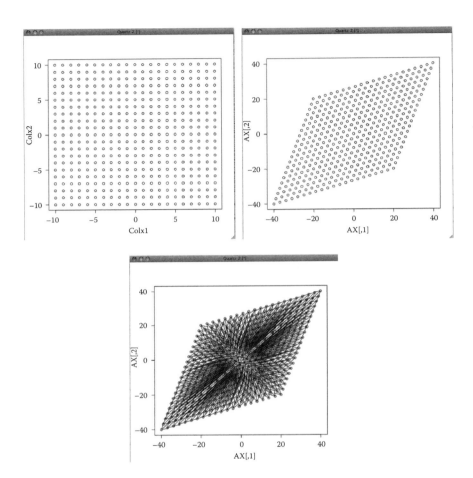

FIGURE 5.6
Equally spaced points in a plane (upper left), space transformed by the matrix **A** (upper right), and directions of data movement (bottom).

```
> colx1 <- rep(1:21,21)
> colx2 <- rep(1:21,rep(21,21))
> X <- rbind (colx1, colx2)
> X <- X -11
> X <- t(X)
> plot(X)
```

The result is shown in the top left panel of Figure 5.6.
   We then define matrix **A**. Confirm that **A** is defined as intended.

```
> A <- c(3,1,1,3)
> A <- matrix (A, nrow=2)
> A
```

```
        [,1] [,2]
[1,]       3    1
[2,]       1    3
```

Observe how **A** transforms the space.

```
> counter <- 1:441
> AX <- rbind(c(0,0), c(counter))
> for (i in 1:441) {AX[,i] <- A %*% X[i,]}
> AX <- t(AX)
> plot(AX)
```

The points originally scattered in a square space are now scattered in a dia-mond-shaped space. This transformation is referred to as a projection by the matrix **A** and is shown in the top right panel of Figure 5.5.

Observe how the original data are transformed into the projected data. The *arrows* command shows the movement between two data. The *arrows* command requires columns as parameters; here, we specify the first and second columns in the original and projected data.

```
> arrows(X[,1],X[,2],AX[,1],AX[,2])
```

The interesting result is shown at the bottom of Figure 5.5. Most arrows are rotated to some degree. A sequence of arrows winds to make curves. Two constant directions exist in that the head of one arrow overlaps the tail of the next one. One direction is in the longer diagonal line of the rhombus from the bottom left to the top right, and the other is in the shorter diagonal line from the top left to bottom right. These directions stand out like animal trails that start at the origin and push through the grass in both directions.

Mathematically, data points move in some constant directions; that is, the movements in these directions are some scalar multiple of the origi-nals. Although most directions change when multiplying by a matrix, some invariable directions always exist. In this example, there are two such direc-tions; the movements along the longer diagonal line are magnified more than movements along the shorter diagonal line.

In general, we denote a vector that does not move as l, and a scalar multiple as λ. Then, the data in the invariant direction when projected by the matrix **A** build a λ-times vector; we can describe the relationship as $\mathbf{A}l = \lambda l$.

This relationship is formalized in the following definition.

There is a relationship where **A** is a square matrix such that

$$\mathbf{A}l = \lambda l \qquad (l \neq \mathbf{o}) \qquad (5.20)$$

The condition $l \neq \mathbf{o}$ means that l is not the null vector $\mathbf{o}$ $\left(= \begin{bmatrix} 0 \\ 0 \end{bmatrix}\right)$.

The vector l is an eigenvector and the scalar λ is an eigenvalue. We can transform Equation (5.20) to solve it as follows:

$$Al - \lambda l = o$$

The *I* is a unit matrix,

$$\begin{bmatrix} 1 & 0 \\ 0 & 1 \end{bmatrix}$$

in the case of two dimensions.

To confirm,

$$\begin{bmatrix} 1 & 0 \\ 0 & 1 \end{bmatrix} \begin{bmatrix} a & b \\ c & d \end{bmatrix} = \begin{bmatrix} 1a+0c & 1b+0d \\ 0a+1c & 0b+1d \end{bmatrix} = \begin{bmatrix} a & b \\ c & d \end{bmatrix}$$

$$(A - \lambda I)l = o \tag{5.21}$$

Here is another relationship called an eigen-equation. The eigenvalue λ of matrix **A** is obtained by solving this eigen-equation.

$$|A - \lambda I| = 0 \tag{5.22}$$

The symbol $|\ |$ indicates a determinant. Calculate the eigenvalue of matrix

$$A = \begin{bmatrix} 3 & 1 \\ 1 & 3 \end{bmatrix}$$

using the equation above.

$$|A - \lambda I| = \begin{bmatrix} 3 & 1 \\ 1 & 3 \end{bmatrix} - \lambda \begin{bmatrix} 1 & 0 \\ 0 & 1 \end{bmatrix} = \begin{vmatrix} 3-\lambda & 1 \\ 1 & 3-\lambda \end{vmatrix}$$

$$= (3-\lambda)^2 - 1^2 \tag{5.23}$$

$$= (3-\lambda+1)(3-\lambda-1) = (4-\lambda)(2-\lambda) = 0$$

The results for λ are 4 and 2. That is, a direction invariant vector is magnified by a factor of 4 along the longer diagonal line of the rhombus, and another direction invariant vector is magnified by a factor of 2 along the shorter diagonal line.

Then, we find the eigenvectors using Equation (5.21).

Let the first eigenvector along the longer diagonal line of the rhombus be

$$l_1 = \begin{bmatrix} l_1 \\ l_2 \end{bmatrix}$$

and the corresponding eigenvalue be λ_1. We have already obtained $\lambda_1=4$ above.

$$(A - \lambda_1 I)l_1 = \left(\begin{bmatrix} 3 & 1 \\ 1 & 3 \end{bmatrix} - 4 \begin{bmatrix} 1 & 0 \\ 0 & 1 \end{bmatrix} \right) \begin{bmatrix} l_1 \\ l_2 \end{bmatrix}$$

$$= \begin{bmatrix} 3-4 & 1-0 \\ 1-0 & 3-4 \end{bmatrix} \begin{bmatrix} l_1 \\ l_2 \end{bmatrix} = \begin{bmatrix} -1 & 1 \\ 1 & -1 \end{bmatrix} \begin{bmatrix} l_1 \\ l_2 \end{bmatrix} = \begin{bmatrix} -l_1 + l_2 \\ l_1 - l_2 \end{bmatrix} = \begin{bmatrix} 0 \\ 0 \end{bmatrix}$$

(5.24)

Because the right side of Equation (5.21) is $\begin{bmatrix} 0 \\ 0 \end{bmatrix}$, we obtain $-l_1 + l_2 = 0$ and $l_1 - l_2 = 0$. Both of these mean that $l_1 = l_2$, that is, any value is acceptable for both l_1 and l_2 simultaneously. The C is some arbitrary constant value other than 0.

$$\mathbf{l}_1 = \begin{bmatrix} l_1 \\ l_2 \end{bmatrix} = C \begin{bmatrix} 1 \\ 1 \end{bmatrix}$$

(5.25)

We have obtained the first eigenvector in this way. The length of an eigenvector is generally 1 and is referred to as a normalized eigenvector. The length of a vector is the square root of the sum of the squares of each element. Thus, we can find the constant that makes the length of an eigenvector equal to unity as follows:

$$C\sqrt{l_1^2 + l_2^2} = 1$$

$$C = \frac{1}{\sqrt{l_1^2 + l_2^2}}$$

(5.26)

$$C = \frac{1}{\sqrt{2}}$$

The first eigenvector is thus

$$\begin{bmatrix} \dfrac{1}{\sqrt{2}} \\ \dfrac{1}{\sqrt{2}} \end{bmatrix}$$

The second eigenvector $\mathbf{l}_2$ is also calculated as above:

$$
\begin{aligned}
\left(\mathbf{A}=\lambda_2 I\right)\mathbf{l}_2 &= \left(\begin{bmatrix} 3 & 1 \\ 1 & 3 \end{bmatrix} - 2\begin{bmatrix} 1 & 0 \\ 0 & 1 \end{bmatrix}\right)\begin{bmatrix} l_1 \\ l_2 \end{bmatrix} \\
&= \begin{bmatrix} 3-2 & 1-0 \\ 1-0 & 3-2 \end{bmatrix}\begin{bmatrix} l_1 \\ l_2 \end{bmatrix} = \begin{bmatrix} 1 & 1 \\ 1 & 1 \end{bmatrix}\begin{bmatrix} l_1 \\ l_2 \end{bmatrix} = \begin{bmatrix} l_1+l_2 \\ l_1+l_2 \end{bmatrix} = \begin{bmatrix} 0 \\ 0 \end{bmatrix}
\end{aligned}
\tag{5.27}
$$

Now, we have $l_1 + l_2 = 0$, i.e., $l_1 = -l_2$.

$$
\mathbf{l}_2 = \begin{bmatrix} l_1 \\ l_2 \end{bmatrix} = C\begin{bmatrix} 1 \\ -1 \end{bmatrix}
\tag{5.28}
$$

Because the square of –1 is 1, Equation (5.26) can be applied directly. The second eigenvector is calculated to be

$$
\begin{bmatrix} \dfrac{1}{\sqrt{2}} \\ -\dfrac{1}{\sqrt{2}} \end{bmatrix}
$$

Eigenvalues and eigenvectors of a square matrix $\mathbf{A}$ are solved by the procedure described above. Because this is too time-consuming to solve manually for more than three dimensions, a computational method, such as the Jacobian method using a computer, is generally used.

In the case where $\mathbf{A}$ is a symmetrical matrix (i.e., where the elements are symmetrical with respect to the diagonal elements running from top left to bottom right), its eigenvectors lie at right angles to each other. Because we use a variance of covariance matrix or correlation matrix for $\mathbf{A}$ in PCA, the eigenvectors lie at right angles to each other in all cases. Also, in the example given above, one eigenvector points to the upper right and the other points to the lower right; they lie at an angle of 90° to each other.

Another characteristic is that the sum of eigenvectors equals the trace of matrix $\mathbf{A}$. The trace is the sum of the diagonal elements of a matrix, and it is represented as $tr\mathbf{A}$. In the example above, the sum of eigenvalues is $4 + 2 = 6$, which is equal to $tr\mathbf{A}$,

$$
tr\mathbf{A}\begin{bmatrix} 3 & 1 \\ 1 & 3 \end{bmatrix} = 3+3 = 6
\tag{5.29}
$$

5.1.5 Explanation with Kansei Evaluation Data

5.1.5.1 Beer Can Data

Let us calculate the two-dimensional case using actual Kansei engineering data obtained by measurement. Table 5.1 shows a portion of actual evaluation data for 56 beer cans with different designs. Each evaluation value is averaged among eight participants. We can see the Kansei words *high grade* and *ambience* are highly correlated.

TABLE 5.1

Averaged Evaluation Data for Beer Cans

	Name of Beer	High Grade	Ambience		Name of Beer	High Grade	Ambience
1	Mack	2.75	3.88	29	Pig's Eye	2.38	2.88
2	Koff Red	3.88	3.75	30	Heineken	3.25	3.13
3	Olvi	3.63	3.38	31	Cobra	1.88	1.88
4	Coors Light	3.63	3.25	32	Dressler	1.75	2.13
5	Stroh's NA	4.25	3.75	33	Red Wolf	3.63	3.63
6	Carlsberg	2.75	3.38	34	Old Milwaukee	4.38	3.50
7	Texas Select	3.25	3.25	35	Labatt Blue	2.75	3.50
8	Cass White	2.38	2.25	36	West End	2.75	3.50
9	Schlitz Blue Ox	3.38	3.25	37	Clausthaller	4.13	4.13
10	Saku	4.50	4.25	38	Buckler	2.75	3.25
11	Murphy's	3.25	2.75	39	Young's London Lager	2.00	2.13
12	Tiger	3.63	3.50	40	Royal Dutch	3.63	2.75
13	Newquay	3.75	4.13	41	Red Bull	2.00	2.25
14	Miller	3.50	3.38	42	Lapin Kulta	3.50	3.63
15	Swan Light	2.50	2.63	43	Karjala	4.38	4.00
16	Carlsberg Special	3.00	3.00	44	Belgium Brown	3.63	4.00
17	Belgian Gold	3.88	3.50	45	Old Milwaukee NA	4.00	3.63
18	Bass Pale Ale	2.50	2.75	46	Heineken Dark	3.88	3.63
19	Karhu	2.50	2.50	47	Tuborg	4.75	4.38
20	Brahma	3.75	3.13	48	Hite	2.00	1.75
21	Michelob Golden	4.00	3.63	49	Budweiser	3.63	3.25
22	Staropramen	3.88	4.00	50	Schaefer Light	2.38	3.13
23	Lowenbrau White	4.00	4.25	51	Brewry	3.38	3.13
24	Piels Red	3.63	3.50	52	Koff Black	4.13	4.13
25	Malibu	1.88	2.88	53	Lowenbrau Blue	2.63	2.75
26	Whitbread Pale Ale	3.38	3.13	54	Miller Lite	3.00	3.25
27	Cass Blue	2.63	2.88	55	Stroh's Deep	4.38	4.00
28	Kaisordom	4.25	4.00	56	Coors Gold	4.13	3.38

5.1.5.2 *Calculation of Variance and Covariance*

With the help of R, let us calculate variance and covariance.

```
> HighGrade <-
c(2.75,3.875,3.625,3.625,4.25,2.75,3.25,2.375,3.375,4.5,3.25,3
.625,3.75,3.5,2.5,3,3.875,2.5,2.5,3.75,4,3.875,4,3.625,1.875,3
.375,2.625,4.25,2.375,3.25,1.875,1.75,3.625,4.375,2.75,3.5,4.1
25,2.75,2,3.625,2,3.5,4.375,3.625,4,3.875,4.75,2,3.625,2.375,3
.375,4.125,2.625,3,4.375,4.125)
> Ambience <-
c(3.875,3.75,3.375,3.25,3.75,3.375,3.25,2.25,3.25,4.25,2.75,3.
5,4.125,3.375,2.625,3,3.5,2.75,2.5,3.125,3.625,4,4.25,3.5,2.87
5,3.125,2.875,4,2.875,3.125,1.875,2.125,3.625,3.5,3.5,2.75,4.1
25,3.25,2.125,2.75,2.25,3.625,4,4,3.625,3.625,4.375,1.75,3.25,
3.125,3.125,4.125,2.75,3.25,4,3.375)
> plot(HighGrade,Ambience)
```

The variables `HighGrade` and `Ambience` are vectors. We build a matrix by combining them in columns.

```
> mat <- cbind(HighGrade, Ambience)
> mat
       HighGrade Ambience
  [1,]     2.750    3.875
  [2,]     3.875    3.750
            :        :
 [56,]     4.125    3.375
```

We can calculate the variance and covariance matrix immediately using the var function.

```
> A <- var(mat)
> A
           HighGrade   Ambience
HighGrade  0.6061688  0.4076705
Ambience   0.4076705  0.3950284
```

Plug this matrix into Equation (5.23) and round each value off to two decimal places to simplify calculation.

$$|\mathbf{A} - \lambda I| = \left| \begin{bmatrix} 0.61 & 0.41 \\ 0.41 & 0.40 \end{bmatrix} - \lambda \begin{bmatrix} 1 & 0 \\ 0 & 1 \end{bmatrix} \right| = \left| \begin{matrix} 0.61 - \lambda & 0.41 \\ 0.41 & 0.40 - \lambda \end{matrix} \right|$$

$$= \begin{bmatrix} 0.61 - 0.93 & 0.41 - 0 \\ 0.41 - 0 & 0.40 - 0.93 \end{bmatrix} \begin{bmatrix} l_1 \\ l_2 \end{bmatrix} = \begin{bmatrix} -0.32 & 0.41 \\ 0.41 & -0.53 \end{bmatrix} \begin{bmatrix} l_1 \\ l_2 \end{bmatrix}$$

(5.30)

$$= \begin{bmatrix} -0.32l_1 + 0.41l_2 \\ 0.41l_1 - 0.53l_2 \end{bmatrix} = \begin{bmatrix} 0 \\ 0 \end{bmatrix}$$

The first eigenvalue is 0.93 and the second eigenvalue is 0.08 using the solution formula.

The first eigenvector is calculated using Equation (5.24).

$$
\begin{aligned}
(A = \lambda_1 I) l_1 &= \left(\begin{bmatrix} 0.61 & 0.41 \\ 0.41 & 0.40 \end{bmatrix} - 0.93 \begin{bmatrix} 1 & 0 \\ 0 & 1 \end{bmatrix} \right) \begin{bmatrix} l_1 \\ l_2 \end{bmatrix} \\
&= \begin{bmatrix} 0.61 - 0.93 & 0.41 - 0 \\ 0.41 - 0 & 0.40 - 0.93 \end{bmatrix} \begin{bmatrix} l_1 \\ l_2 \end{bmatrix} = \begin{bmatrix} -0.32 & 0.41 \\ 0.41 & -0.53 \end{bmatrix} \begin{bmatrix} l_1 \\ l_2 \end{bmatrix} \quad (5.31) \\
&= \begin{bmatrix} -0.32l_1 + 0.41l_2 \\ 0.41l_1 = 0.53l_2 \end{bmatrix} = \begin{bmatrix} 0 \\ 0 \end{bmatrix}
\end{aligned}
$$

We obtain $l_1 = 1.28l_2$ which gives

$$C \begin{bmatrix} 1.28 \\ 1 \end{bmatrix}$$

We normalize the eigenvector by assigning $C\sqrt{1.28^2 + 1^2} = 1$. Thus the constant C is 0.62 from Equation (5.26) and the first (normalized) eigenvector is

$$\begin{bmatrix} 0.79 \\ 0.62 \end{bmatrix}$$

We can obtain the second eigenvector in the same way by substituting -0.08 for -0.93 in the above equation.

The R software package provides the `eigen` function for calculating eigenvalues and eigenvectors. The results may be slightly different from the values calculated above because we rounded the data to two decimal places.

```
> eigenresult <- eigen(A)
> eigenresult
$values
[1] 0.92171650 0.07948074

$vectors
            [,1]        [,2]
[1,] 0.7907877  0.6120905
[2,] 0.6120905 -0.7907877
```

The formulas for z built by the first and the second eigenvalues are called the first principal component and the second principal component, respectively. The formulas for the first principal component z_1 and the second principal component z_2 are

$$z_1 = 0.79y_1 + 0.61\ y_2$$

$$z_2 = 0.61y_1 - 0.79\ y_2 \tag{5.32}$$

5.1.5.3 *Plot of the Results*

Here, we finish the general explanation of principal components, eigenvalues, and eigenvectors. Finding the maximum variance means finding the eigenvalues and eigenvectors using equations or by any other method. Let us examine how they correspond to original evaluation data.

The left part of Figure 5.7 is a plot of the original data, drawn by entering plot(HighGrade,Ambience). The right part shows how the 20×20 points placed on a plane in the previous section were transformed by applying a matrix **A**. A variance-of-covariance matrix transforms the space so it is pulled in directions to ensure that the original data have maximum variance. The

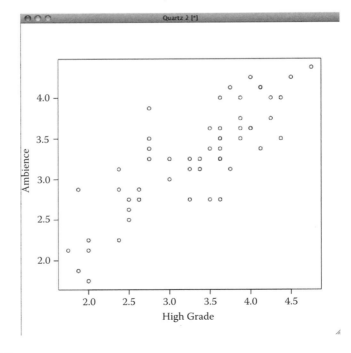

FIGURE 5.7
Relationship between *high grade* and *ambience*.

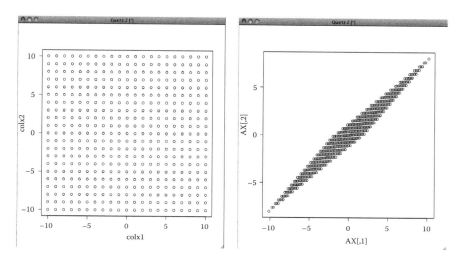

FIGURE 5.8
Plot of the original data (left) and transformation by a matrix **A** (right).

magnifying power changes with the pulling strength. Thus the first eigen-vector of the variance-of-covariance matrix is found where the maximum variance of the original data occurs. PCA represents the directions in which the space is pulled as eigenvectors and the pulling strengths as eigenvalues.

```
> colx1 <- rep(1:21,21)
> colx2 <- rep(1:21,rep(21,21))
> X <- rbind (colx1, colx2)
> X <- X -11
> X <- t(X)
> plot(X)
> counter <- 1:441
> AX <- rbind(c(0,0), c(counter))
> for (i in 1:441) {AX[,i] <- A %*% X[i,]}
> AX <- t(AX)
> plot(AX)
```

5.1.5.4 Contribution Ratio

The contribution ratio is an index of how much each principal component represents the characteristics of the original data. An eigenvalue correspond-ing to each principal component is used as this index. As mentioned above, the larger the variance along an eigenvector, the larger the eigenvalue will be in proportion to the variance. Thus, the larger the eigenvalue, the more the corresponding principal component represents the characteristics of the original data.

We mentioned that the sum of eigenvalues is equal to the trace of matrix $\mathbf{A}$ ($tr\mathbf{A}$). Thus, the ratio of an eigenvalue component to $tr\mathbf{A}$ represents how much the corresponding principal component contributes to describing the characteristics of the original data. This ratio is called the contribution ratio.

In this case, the contribution ratio of the first principal component is obtained by dividing the first eigenvalue by $tr\mathbf{A}$ (see Equation 5.30):

$$\frac{0.93}{0.61+0.4}=0.92 \tag{5.33}$$

The contribution ratio is thus 92%. We can consider that the first principal component, a synthetic variable, represents by itself much of the information that was originally represented by a combination of two variables.

The cumulative contribution ratio is another index that is also commonly used. This is a summation of the contribution ratios corresponding to the first through the nth principal component. In our example, the contribution ratio of the second principal component is calculated to be $0.08/(0.61 + 0.4) = 0.08$, or 8%. Then, the cumulative contribution ratio up to the second principal component becomes 100%. Because the original data are two-dimensional, this confirms that the two principal components together represent all the characteristics of the original data.

5.1.5.5 Principal Component Scores

The principal score of a sample indicates where the sample is placed along the principal component. The principal score is obtained by substituting data into Equation (5.32), where we used the average deviation (i.e., the data with the average subtracted from it).

R calculates the average value with the function mean.

```
> mean(HighGrade)
[1] 3.321429
> mean(Ambience)
[1] 3.28125
```

We can obtain a principal component score of the 10th beer can sample, SAKU, along the first principal component by

$$z_1 = 0.79(4.50 - 3.321) + 0.61(4.25 - 3.281) = 1.5225$$

We can also obtain the second principal component score by

$$z_1 = 0.79(2.75 - 3.321) + 0.61(3.875 - 3.281) = -0.0887$$

FIGURE 5.9
Saku (left) and Mack (right).

The first principal component score for the first beer can sample Mack can be obtained by

$$z_1 = 0.79(2.75 - 3.321) + 0.61(3.875 - 3.281) = -0.0887$$

We can express the above results as follows: The first principal component that summarizes the Kansei words *high grade* and *ambience* can be expressed as *degree of gorgeous*. Saku is judged to be very gorgeous, whereas Mack is not so gorgeous. Saku is an Estonian beer, the can painted a beautiful, metallic deep blue. The Norwegian Mack beer can is also metallic blue, but the slight difference in the shade of blue and the design of the label seem to affect the whole impression (Figure 5.9).

We can calculate the principal component scores in one operation using this procedure:

```
> scaled.mat <- scale(mat, center=TRUE, scale=FALSE)
> scaled.mat
         HighGrade  Ambience
 [1,]  -0.57142857   0.59375
 [2,]   0.55357143   0.46875
        :             :
```

When we apply the function scale to a matrix, it normalizes the elements in each column of the matrix. If we use center=TRUE as shown above, the elements in each column become the average deviation by subtracting the center (average). Another parameter, `scale`, makes the variance 1; however, we do not use it here, so we set it to `FALSE`.

Here, we solve the first principal component scores for all samples. The [,1] denotes a vector that is built based on its first column.

```
> pcscore <- scaled.mat %*% eigenresult$vectors[,1]
> pcscore
              [,1]
  [1,]  -0.08844995
  [2,]   0.72467491
  [3,]   0.29744404
  [4,]   0.22093273
  [5,]   1.02122030
  [6,]  -0.39449521
  [7,]  -0.07561266
  [8,]  -1.37964243
  [9,]   0.02323580
 [10,]   1.52496249
   :
```

Let us sort the principal component scores we have obtained. The parameter setting `index.return=TRUE` shows the data in the original order.

```
> sort(pcscore, index.return=TRUE)
$x
 [1] -2.00457022 -1.98223307 -1.95039606 -1.75269913
-1.67618782 -1.39247971
 [7] -1.37964243 -1.12777133 -1.05126002 -0.99708585
-0.97474870 -0.87590024
   :

$ix
 [1] 31 48 32 39 41 25  8 19 15 29 18 53 50 27 38 16  6 11 35
54 36 30  1 40
[25]  7 26 51  9 14  4 49 20  3 42 12 24 33 17 46 44 56  2 21
45 13 22 34  5
[49] 23 37 52 28 43 55 10 47
```

We observe that the least gorgeous sample is number 31 (Cobra), and the most gorgeous one is number 47 (Tuborg). The former was evaluated low in both *high grade* and *ambience,* and the latter was evaluated high in both Kansei words (Figure 5.10).

5.1.5.6 *Principal Component Loadings*

A principal component loading is a correlation between the principal component score and the original variable. In other words, it represents how

FIGURE 5.10
Cobra (left) and Tuborg (right).

original variables correlate with the principal component, that is, synthesized variables.

You can calculate the principal loading for a Kansei word as follows. The
function cor shows the correlation coefficient between two vectors.

```
> cor(pcscore, HighGrade)
          [,1]
[1,] 0.9751284
> cor(pcscore, Ambience)
          [,1]
[1,] 0.9349754
```

The principal component loading of the Kansei term *high grade* along the first
principal component is 0.97, and that of *ambience* is 0.93. We can see that both
words have great effects on the first principal component.

We can calculate the correlations for several Kansei words (two, in this
example) simultaneously. We define a function cat.floadings so that it
calculates correlation coefficients between a column of the original data
(mat) and pcscore, for each column using the apply function. The second
parameter for apply means that the specified calculation is performed in
a column.

```
> cal.floadings <- function(vect1,vect2) cor(vect1,vect2)
> apply(mat,2,cal.floadings, pcscore)
HighGrade  Ambience
0.9751284 0.9349754
```

5.1.5.7 Example Analysis of Actual Kansei Evaluation Data

Let us analyze real data acquired from a Kansei evaluation of many objective samples using many Kansei words. We first perform the PCA, and then we interpret results. The file beerAveragedSmallSet.txt is a limited one from evaluation data of 56 beer cans with 55 Kansei words, reduced from 92.

Make sure to set R's working directory to the folder that contains the data.

```
> Beer <- read.table("beerAveragedSmallSet.txt", header=T,
row.names=1)
> var(Beer)
```

#The "row.names=1" option means the first row of the read data was assigned as the labels of the rows.

We build a variance-covariance matrix from the matrix Beer.

```
> eigenresult <- eigen(var(Beer))
> eigenresult
> plot(eigenresult$values)
```

Figure 5.12 shows a plot of the eigenvalues obtained. This plot is often called a *scree plot*. The values of first, second, and subsequent eigenvectors are found from left to right along the horizontal axis of the graph. This helps determine how many eigenvectors are important for understanding the structure of the data. In this case, Figure 5.12 shows that the slope becomes quite gentle after the fourth eigenvector. This means that the first four eigenvectors are sufficient to characterize all the information. Next, we compute the principal component scores and principal component loadings corresponding to the first four eigenvectors.

◆	A	B	C	D	E
1	BeerCanName	HighGrade	Ambience	Beautiful	Homely
2	Mack	2.75	3.875	3.875	2
3	KOFFred	3.875	3.75	3.375	1.625
4	OLVI	3.625	3.375	2.75	3.75
5	CoorsLIGHT	3.625	3.25	3.5	3.375
6	STROHsNA	4.25	3.75	3.375	1.625
7	Carlsberg	2.75	3.375	3.875	3.25
8	TexasSelect	3.25	3.25	3.5	2.375
9	CASSwhite	2.375	2.25	3	3
10	SchulitzBlueOx	3.375	3.25	3.25	2.5
11	SAKU	4.5	4.25	4.5	1.875
12	MURPHYs	3.25	2.75	2.375	4
13	TIGER	3.625	3.5	2.875	3.25
14	NEWQUAY	3.75	4.125	4	2.75
15	Miller	3.5	3.375	3.75	2

FIGURE 5.11
Standard format for Kansei data analysis (averaged between subjects).

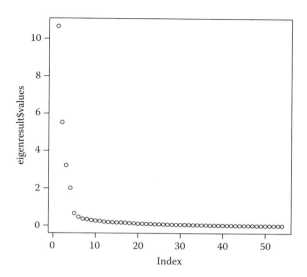

FIGURE 5.12
Plot of the eigenvalues.

```
> centered.Beer <- scale(Beer, center=TRUE, scale=FALSE)
> pcscore1 <- centered.Beer %*% eigenresult$vectors[,1]
> pcscore2 <- centered.Beer %*% eigenresult$vectors[,2]
> pcscore3 <- centered.Beer %*% eigenresult$vectors[,3]
> pcscore4 <- centered.Beer %*% eigenresult$vectors[,4]
> cal.floadings <- function(vect1,vect2) cor(vect1,vect2)
> apply(Beer, 2, cal.floadings, pcscore1)
    HighGrade        Ambience        Beautiful          Homely
Coarse          Light          Smart
   0.06700292     -0.05361434     -0.60080434      0.12034522
0.58623048     -0.89365390     -0.82468884
        Slim            Sweet            Young          Simple
Soft        Elegant          Modern
  -0.80348727     -0.79383482     -0.72004462     -0.49016000
-0.85643850     -0.25073901     -0.35059905
        Unique            Lite            Warm           Fresh
Natural         Plain          Active
   0.10877253     -0.89591535      0.03567316     -0.88269452
-0.69234528     -0.85173448      0.18783722
        Steady           Casual           Pretty        Gorgeous
Chilly          Adult      Individual
   0.82064646     -0.72034436     -0.76631202     -0.17664428
-0.76858718      0.49557699      0.66550448
        Gravely         Childish       Monotonous         Refined
Calm       Original          Showy
   0.94228329     -0.62464563     -0.23401084     -0.37498592
-0.25144360      0.48638658      0.18406573
```

```
        Dark          Healthy          Sporty           Bright
Polished        Affected          Hard
   0.68049561    -0.65972122    -0.49294096    -0.66082397
-0.37012898     0.04050878     0.71261433
        Lively OverDecorated     Intelligent         Chic
Diversified     Refreshing        Massive
  -0.28517087     0.67180125    -0.22674194     0.37597821
0.30584656    -0.90270809     0.69268906
   Uninhibited          Sharp       Masculine        Feminine
Cool
  -0.52025101    -0.04834846     0.71996785    -0.73560672
-0.36993940
> floadingsTable <- rbind(apply(Beer, 2, cal.floadings,
pcscore1),apply(Beer, 2, cal.floadings, pcscore2),apply(Beer,
2, cal.floadings, pcscore3),apply(Beer, 2, cal.floadings,
pcscore4))
```

We bind a row of principal component loadings, from the first to the fourth, one by one, to build the matrix floadingsTable. When we input a command longer than one line, a + sign appears automatically on the R console to show that the line follows on from the previous line. We do not have to input the + sign.

```
> floadingsTable
         High-Grade        Ambience  Beautiful          Homely
Coarse    Light        Smart        Slim
[1,]   0.06700292 -0.05361434 -0.6008043  0.120345217 0.5862305
-0.89365390 -0.82468884 -0.80348727
[2,]   0.05997763  0.04676011  0.1886993 -0.905944389 0.2767533
0.03524990 -0.29499829 -0.33465552
[3,]  -0.82111913 -0.63023965 -0.3872559 -0.004852964 0.3362297
0.20750499  0.06505798 -0.02458890
[4,]  -0.49623949 -0.57428869 -0.5017368  0.163939104 0.3475123
-0.05537237 -0.24939617 -0.23337282
    :        :
> t(floadingsTable)
```

	[,1]	[,2]	[,3]	[,4]
HighGrade	0.06700292	0.05997763	**-0.821119133**	-0.49623949
Ambience	-0.05361434	0.04676011	**-0.630239650**	**-0.57428869**
Beautiful	**-0.60080434**	0.18869931	-0.387255866	-0.50173676
Homely	0.12034522	**-0.90594439**	-0.004852964	0.16393910
Coarse	**0.58623048**	0.27675326	0.336229722	0.34751230
Light	**-0.89365390**	0.03524990	0.207504994	0.05537237
Smart	**-0.82468884**	-0.29499829	0.065057979	-0.24939617
Slim	**-0.80348727**	-0.33465552	-0.024588897	-0.23337282
Sweet	**-0.79383482**	0.02621867	-0.390455029	0.19926599
Young	**-0.72004462**	0.34849869	0.343264675	-0.07649163
Simple	-0.49016000	**-0.74289345**	0.107577538	-0.05900914

Soft	**-0.85643850**	-0.27215841	-0.204582890	0.08506014
Elegant	-0.25073901	0.16149695	**-0.812826026**	-0.29619013
Modern	-0.35059905	0.04496325	-0.457227294	-0.21607353
Unique	0.10877253	**0.60396500**	0.253651060	0.05524448
Lite	**-0.89591535**	0.20079443	0.069783545	-0.05243363
Warm	0.03567316	0.28521599	**-0.598931303**	0.48310204
Fresh	**-0.88269452**	0.18526219	0.249081350	-0.10180523
Natural	**-0.69234528**	-0.28102791	-0.101152428	0.06685158
Plain	**-0.85173448**	-0.19838497	0.251992872	-0.08823566
Active	0.18783722	**0.81376640**	0.343151758	-0.07064497
Steady	**0.82064646**	-0.20130478	0.048870107	-0.26751064
Casual	**-0.72034436**	0.24050567	0.110498755	0.08272009
Pretty	**-0.76631202**	0.11130386	-0.259826864	0.28138309
Gorgeous	-0.17664428	**0.71245897**	-0.497455391	-0.17366665
Chilly	**-0.76858718**	-0.10389260	0.388854041	-0.39941178
Adult	0.49557699	-0.27808127	-0.485725023	-0.46532038
Individual	**0.66550448**	0.38205823	0.064425039	-0.01780874
Gravely	**0.94228329**	-0.12511179	-0.034592438	-0.01894291
Childish	**-0.62464563**	0.14368304	0.312722086	0.47505025
Monotonous	-0.23401084	**-0.65445817**	0.165728525	0.24865249
Refined	-0.37498592	0.14715214	-0.238914027	**-0.56975074**
Calm	-0.25144360	**-0.79529718**	-0.367500102	-0.18536605
Original	0.48638658	**0.70759754**	0.147080135	-0.24443996
Showy	0.18406573	**0.93700503**	0.002186515	-0.02826798
Dark	**0.68049561**	**-0.65494294**	0.077248197	0.00963828
Healthy	**-0.65972122**	0.44062046	0.284715060	0.07767756
Sporty	-0.49294096	0.43235333	**0.522137460**	-0.20965587
Bright	**-0.66082397**	**0.63216682**	-0.212257668	0.15612149
Polished	-0.37012898	**0.75962106**	-0.124107664	-0.05767164
Affected	0.04050878	0.46743844	**-0.557799731**	-0.34437457
Hard	**0.71261433**	0.35707307	0.176984764	-0.34202882
Lively	-0.28517087	**0.71309154**	0.378837029	-0.19901520
Overdecorated	0.67180125	0.56004697	-0.023637352	-0.04597013
Intelligent	-0.22674194	-0.19621633	**-0.624941139**	-0.45193469
Chic	0.37597821	**-0.66769561**	-0.189103186	-0.37388238
Diversified	0.30584656	**0.64249637**	0.118788310	-0.20782494
Refreshing	**-0.90270809**	-0.04569923	0.193756480	-0.13184462
Massive	**0.69268906**	0.44295127	-0.157155924	-0.14608327
Uninhibited	**-0.52025101**	-0.40183442	0.152619540	-0.07857898
Sharp	-0.04834846	-0.14392199	**0.570691845**	**-0.63378557**
Masculine	**0.71996785**	0.01449282	0.464481904	-0.34003002
Feminine	**-0.73560672**	0.12016334	-0.525212929	0.24310267
Cool	-0.36993940	-0.17093167	**0.520491001**	**-0.66495736**

>

The function t() was used to transpose rows and columns of the table to make it easier to interpret. Principal component loadings shown in boldface are 0.5 or larger. Let us further examine this table of principal component loadings.

The words *hard, gravely, masculine,* and *massive* received large positive loading values along the first principal component. On the other hand, *light, soft, refreshing,* and *feminine* received large negative values. We interpret this principal component as the impression of lightness or heaviness.

Along the second principal component, the words *active, showy, gorgeous, novel, sophisticated,* and *lively* received large positive values. Large negative values were assigned to *plain, simple,* and *calm* along the same principal component. We interpret this principal component as activity, or the degree of showiness or plainness.

Sharp, sporty, and *cool* received large positive values, while *high-grade, elegant, ambience,* and *intellectual* received large negative values along the third principal component. We interpret this as the degree of quality.

No word received positive values larger than 0.5 along the fourth principal component, although *warm* received 0.48 and *childish* received 0.47. *Sharp, cool,* and *refined* had large negative values. This principal component must be related to the third principal component, because the two principal components share some words.

These interpretations of PCA results are done in the traditional way. We can go further with graphical visualization with R.

```
> floadingsTable<-t(floadingsTable)
```

Use the function "row.names()" to assign the following Kansei words to a row vector.

```
> row.names(floadingsTable)
[1] "HighGrade"      "Ambience"     "Beautiful"     "Homely"
"Coarse"        "Light"
 [7] "Smart"         "Slim"         "Sweet"         "Young"
"Simple"        "Soft"
[13] "Elegant"       "Modern"       "Unique"        "Lite"
"Warm"          "Fresh"
[19] "Natural"       "Plain"        "Active"        "Steady"
"Casual"        "Pretty"
[25] "Gorgeous"      "Chilly"       "Adult"
"Individual"    "Gravely"      "Childish"
[31] "Monotonous"    "Refined"      "Calm"
"Original"      "Showy"        "Dark"
[37] "Healthy"       "Sporty"       "Bright"
"Polished"      "Affected"     "Hard"
[43] "Lively"        "OverDecorated" "Intelligent"  "Chic"
"Diversified"   "Refreshing"
```

```
[49] "Massive"          "Uninhibited"    "Sharp"
"Masculine"      "Feminine"        "Cool"
> plot(floadingsTable[,1],floadingsTable[,2], ylim=c(-
1,1),xlim=c(-1,1))
```

Kansei words appear in the plot of factor loadings using the one-liner as follows. When you click a dot, the corresponding Kansei word appears. Using drawing software such as Adobe Illustrator, we can see all words.

```
> identify(floadingsTable[,1],floadingsTable[,2],row.
names(floadingsTable))
```

We investigate the structure built by first and second principal component loadings in the plot. In Figure 5.11, we added arrows for interpretation by hand, where some Kansei words are slightly moved so as not to hide others. Along the first PC, opposition between lightness (left) — heaviness (right) is apparent. Along the second PC, activity and showiness (upper) — plainness (lower) is also apparent opposition between them. We can find other structures in addition to PC structures. A polarized structure from upper right to lower left goes through *overdecorated*, *massive*, and *simple* on the opposite side. Another polarization lies from upper left to lower right, that goes through *bright* and *sporty*, and on the opposite side, *dark* and *adult*. We can find useful implications in such computer-generated maps.

Now in order to understand the semantic space by considering principal component loadings, let us analyze the principal component scores to reveal the relationships between the design samples and the principal components (Figure 5.13).

```
> pcscoreTable <- cbind(pcscore1,pcscore2,pcscore3,pcscore4)
> pcscoreTable
                        [,1]          [,2]           [,3]
[,4]
Mack              -1.8717147  0.77648230   2.05116362
-1.39805060
KOFFred            1.4276997  3.38140445  -2.27923851
0.87101655
OLVI               4.7533668 -4.49408990  -0.86981308
-0.76994552
 :       :        :        :        :
> plot(pcscoreTable[,1],pcscoreTable[,2],
ylim=c(-7,7),xlim=c(-7,7))
```

It is recommended to make the PC loadings plot square using options "ylim" and "xlim."

```
> identify(pcscoreTable[,1],pcscoreTable[,2],row.
names(pcscoreTable))
```

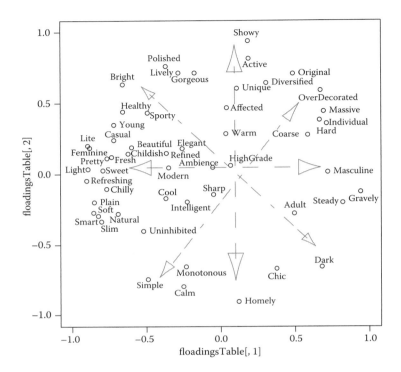

FIGURE 5.13
Plot of principal component loadings (PC1 and PC2).

\# The "identify" command is useful to export the plot with EPS format, when we retouch it using other software (e.g., Adobe Illustrator).

```
> dev.copy2eps ( file = "file_name.eps" )
```

Now observe the scores and loadings along the first principal component (Figure 5.14). Texas Select, Labatt Blue, Old Milwaukee, and many other cans are located on the left, that is, in the direction of the lightness. Many of them are white, silver, or blue. Since many cans are located on the left of the origin, they represent a principal characteristic of beer, that is, lightness and a feeling of refreshment. At the opposite of lightness, *masculine* and *gravely* are mapped and there are placed Red Bull, Red Wolf, Pig's Eye, Schlitz Blue Ox, and Karhu. These cans typically show an illustration of an animal or one-eyed pirate (Pig's Eye). These represent the other principal characteristics of beer strength.

When we go down from the top along the second PC, we see *showy*, *active*, *unique*, and *affected* where Karjala, Brahma, and Coors Gold are placed. They are golden or metallic red. Reversing, when going up from the bottom, we find *homely* and *calm*, with Hite in the corresponding area, which had a large white beige area and unadorned logo at that time of the evaluation. *Showiness*

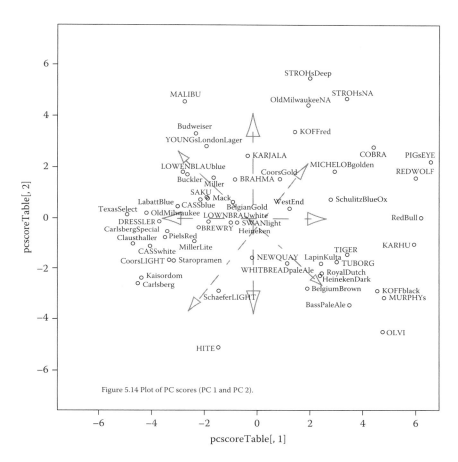

Figure 5.14 Plot of PC scores (PC 1 and PC 2).

FIGURE 5.14
Plot of PC scores (PC1 and PC2).

and *activeness* are also considered important features of many products, because there are many beer cans placed at the corresponding area.

Next, we look at oblique structures. *Overdecorated* and *massive* are in the upper right region of the plot, and there are corresponding beers such as Michelob Golden and Cobra. Michelob has a large black label and white logo on a golden body. Cobra has an illustration of a cobra head covering a whole can. No cans directly correspond to *simple*, the counterpart of them; however, Hite is the closest. Another oblique structure lies from upper left to the lower right. In this direction from the origin to *bright* and *sporty* are Lowenbrau (blue), Buckler, and Miller. Lowenbrau is distinctive light blue. Buckler has a label with fringed yellowish orange; Miller is blue and white and has a large, slanted logo. *Dark* and *adult* are at the opposite end, and there is a group of cans. Most of them are premium beer whose cans have dark colors in common. Olvi has a red label on a pitch-black body. Koff black is colored black, and Murphy's has a very old-style label on a black and beige background. Bass Pale Ale, Belgium Brown, and Heineken dark are dense ale-type beers, and they are commonly colored brown. Royal Dutch, Lapin Kulta, Tuborg, and Tiger are colored with dark, dense tones, although their hues are different.

Computer-generated mappings of PCA help us extract many indispensable aspects of the product. In this section, we see the maps of PC loadings and scores and understand the positions of the products. Additionally, we can find the already well-presented Kansei and one that has not been focused on yet. That kind of information may help us develop a future strategy. Furthermore, combining several different Kansei is a possible approach to develop innovative products.

5.2 Factor Analysis

Factor analysis is often confused with PCA because both are calculated using similar procedures and their results are presented in similar ways. We are frequently asked which analysis should be used. Although these two analyses involve similar calculations, they involve opposing premises.

PCA finds the principal components that are shared by all variables, while factor analysis represents the data with *common factors* that are shared by all variables and *unique factors* that are particular to a variable. The unique factors correspond to a residual of the linear model. Factor analysis does not minimize the residual.

Principal components scores (PCs) are described using the original variables, y, in the model of PCA as follows. In Equation (5.32), for example, we have only two y's (i.e., y_1 and y_2), so we obtain two PCs (i.e., PC_1 and PC_2). Here, suppose that we have n variables as y.

PC_1 = first_eigenvector$_1$ × y_1 + first_eigenvector$_2$ × y_2 + ... + first_eigenvector$_n$ × y_n

PC_2 = second_eigenvector$_1$ × y_1 + second_eigenvector$_2$ × y_2 + ... + second_eigenvector$_n$ × y_n

...

PC_n = nth_eigenvector$_1$ × y_1 + nth_eigenvector$_2$ × y_2 + ... + nth_eigenvector$_n$ × y_n

(5.34)

Conversely, factor analysis is modeled as follows. Unlike PCA, the original variable y is described as a sum of the factor scores of q common factors (assuming that you want q factors to understand the data) and a unique factor. The unique factor of the ith variable y is represented as e_i.

The factor score is solved in a manner similar to the PC, as shown in the following model.

y_1 = first_factor_loading$_1$ × first_factor_score$_1$ + second_factor_loading$_1$ × second_factor_score$_1$ + ... + $q$th_factor_loading$_1$ × $q$th_factor_score$_1$ + e_1

...

y_n = first_factor_loading$_n$ × first_factor_score$_n$ + second_factor_loading$_n$ × second_factor_score+ ... + $q$th_factor_loading$_n$ × $q$th_factor_score$_n$ + e_n

(5.35)

Look at the difference between the two sets of equations. PCA equations have no error terms because the model of PCA incorporates the variances of all variables. In contrast, the model of factor analysis provides error terms to represent factors that differ (are unique) with the variables because factor analysis is used to reproduce the correlation among the original variables using as few factors as possible.

Equation (5.35) is presented in forms of vectors as follows:

$$Y = \Lambda f + e \qquad (5.36)$$

where **Y** is an n-dimensional vector built from the original n variables, **Y** = {$y_1, y_2, ..., y_n$}; **f** is a q-dimensional vector built from q common factors, **f**′ = {$f_1, f_2, ..., f_q$}; **e** is an n-dimensional vector built using n unique factors, **e**′ = {$e_1, e_2, ..., e_n$}; and Λ is an $n × q$-dimensional matrix represented as follows:

$$\Lambda = \begin{bmatrix} \lambda_{11} & \lambda_{12} & \cdots & \lambda_{1q} \\ \lambda_{11} & \lambda_{22} & \cdots & \lambda_{2q} \\ \vdots & \vdots & & \vdots \\ \lambda_{n1} & \lambda_{n2} & \cdots & \lambda_{nq} \end{bmatrix}$$

We set the following conditions. Each common factor, $f_1, f_2, ..., f_q$ of matrix **f**′ has 0 for its mean and 1 for its variance. Each unique factor, $e_1, e_2, ..., e_n$, also has 0 for its mean; their variance is represented as $d_1^2, d_2^2, ..., d_n^2$, respectively.

Let Σ be the variance of a covariance matrix of n-dimensional original variables y. Then, we get Σ as follows:

$$\Sigma = \Lambda \Lambda' + D, \qquad (5.37)$$

where **D** is a matrix of which diagonal elements are $d_1^2, d_2^2, ..., d_n^2$. This equation states that the variance for the covariance matrix of the original variables is the sum of the variance of the factor loadings and the variance of the unique factors. Therefore, your task is to determine the factor loadings and unique factors in the calculation procedure.

Here, we briefly explain communality, a characteristic idea of factor analysis. Communality is the ratio of the part that is represented by the common factors **f** in all variances of the variables. Communality is easier to explain using the following calculation procedure. Many calculation methods can be used for factor analysis; here, we follow the principal factor method.

1. Let **R** be the correlation matrix between the original variables. Then, let **R*** be the matrix in which you set the diagonal elements of **R** as an estimate of communality. Generally, the estimate of communality is set as the square of a multiple correlation coefficient that is obtained by multiple regression analysis, where the objective variable is set as one of y (y_j) and the explanatory variables are set as the rest of all variables y. This is because the y_j that is highly correlated with the other variable y's is expressed with the other y's, such that y_j has high communality.

2. Solve the eigenvalues and eigenvectors of the matrix **R*** as in the case of PCA.

3. Build a q-dimensional vector of factor loadings of the original variables when you want to extract q factors. You can build a column vector of the first factor loadings by putting "SQRT(the first eigenvalue) × the first eigenvector" in the vertical column. Then, build the column vectors until the qth factor loadings, while in a similar way putting "SQRT(the qth eigenvalue) × the qth eigenvector" in the vertical column. Finally, when you collect and put all the factor loading vectors horizontally, you obtain a matrix of the estimate factor loadings $\hat{\Lambda}$ (that expresses the estimate value).

   The principal factor method finishes here because it is not an iterative solution technique. In the case of an iterative solution technique, the next step is as follows.

4. Compare the diagonal elements between **R*** and $\hat{\Lambda} \hat{\Lambda}'$. If the former is sufficiently small, the calculation is finished. Otherwise, replace the diagonal elements of **R*** with $\Sigma \lambda_{jk}^2$, a sum of the q square root of factor loadings attributable to variable j that varies from 1 to q, and return to Step 2.

In conclusion, in comparing PCA and factor analysis, building a factor analysis model is better when you wish to express data with fewer factors. However, exercise caution in a choice of too few factors if you obtain a large value for unique factor loadings.

Many other topics need to be explained for factor analysis, such as rotation of the factor loadings matrix for better understanding and with a stronger algorithm. However, they are beyond the scope of this chapter. For further information, refer to the reference list.

5.3 Cluster Analysis

5.3.1 Objectives of Clustering

When we categorize some things, we can observe the objects with more detail if we obtain some clusters, that is, subsets of objects. Members of a cluster share some attributes; for example, all green plants carry out photosynthesis, or birds and insects that suck honey from trumpet-shaped flowers commonly have long beaks. Detailed study of similarity shared with cluster members guides us to understand whole objects. Thus, clustering is finding a structure in the objects that we know a little about.

Another purpose of clustering is to form the objects' representation and to store them, and communicate with other people. Suppose that we have to memorize 15 different cats. It is easier to memorize general characteristics, for example, they have whiskers, large ears, and tall tails, and characteristics that distinguish them from other cats than to memorize all characteristics of each cat.

In this section, we use following terms: a *cluster* is a set of input objects categorized, a *member* is an input object resulting in a cluster, and *clustering* is to categorize input objects into some clusters.

5.3.2 Use of Cluster Analysis in Kansei Engineering

Cluster analysis is used in Kansei engineering as follows. We first average the Kansei evaluation data among participants. Thus, we have m vectors $\{y_1, y_2, \ldots, y_n\}$, whose elements are averaged evaluation data with n Kansei words, where m is the number of objective samples. Comparison of similarity among m vectors leads to grouping objective samples so that samples with similarly evaluated Kansei words are grouped. We have described in previous sections that we reveal correlational structure of Kansei words using PCA and factor analysis, while we categorize objective evaluation samples using cluster analysis on the same data set.

Samples categorized into a cluster are evaluated similarly on the same set of Kansei words, and they often share some design elements; that is, samples categorized into a cluster with high evaluations on the same Kansei words commonly have an abstract pattern. In such case, we find the characteristic design element has a close relationship with one or more Kansei words.

We often have a huge number of design elements x. In case of direct modeling of relations between x and Kansei words y, such as quantification theory

Type I, it is not practical to take up all x's into the model, because we encounter problems of a limited number of samples or interaction between variables. Therefore, it is useful to perform cluster analysis to obtain important Kansei words and effective design elements in the domain of the objective product, then carefully select the variables to be composed into the model for analysis.

5.3.3 Type of Clustering Methods

We can categorize methods of clustering from various viewpoints. Here, we take a general view from two viewpoints: (1) hierarchical and nonhierarchical methods and (2) process of clustering. Then, a general algorithm of clustering analysis is explained.

5.3.3.1 Hierarchical Methods and Nonhierarchical Methods

The hierarchical method is that samples are merged into a cluster, and then the obtained clusters are merged further into a larger clusters; finally, all clusters are merged into a single cluster that contains all samples. In this way, the obtained clusters are nested or in the hierarchy as shown in Figure 5.15. Some clusters are merged in further steps and others remain alone.

The nonhierarchical method is that any cluster contains other clusters. Clusters are divided into subclusters.

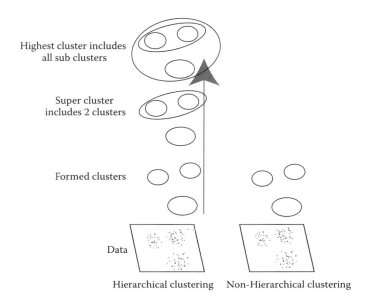

FIGURE 5.15
Hierarchical clustering and nonhierarchical clustering.

We cannot generally compare the advantages of the hierarchical and non-hierarchical methods because they depend on the characteristics of objects and the objectives of clustering. If we categorize samples to distinguish them by their characteristics, hierarchical methods are suitable because they summarize the common information of samples belonging to a cluster at each stage of clustering. In the other cases where the relationships among samples must be expressed straightforwardly, nonhierarchical methods are suitable (Sneath and Sokal, 1973).

In addition, in the case of huge data for more than 100 samples, we often categorize the samples using hierarchical methods first, and then we apply the nonhierarchical method to the obtained clusters.

5.3.3.2 Methods of Hierarchical Clustering

Hierarchical methods build nested clusters. The methods are roughly classified into two types: (1) divisive methods that repeat division of clusters into smaller ones, and (2) agglomerative methods that repeat merging of small clusters finally into a single cluster in all samples included.

Agglomerative methods are often used for hierarchical clustering. We first calculate the similarities or dissimilarities between each two of n samples and build them into an $(n-1) \times (n-1)$ similarity (dissimilarity) matrix. We find a pair of samples with the largest similarity (least dissimilarity) between them and merge them into a new cluster. Then, we rebuild a similarity (dissimilarity) matrix using the new cluster and deleting the two merged samples, which results in a reduced matrix. The new similarities between the new cluster and the others are calculated in various ways, such as an average or a centroid of members in the clusters and the others. We iterate the procedure until we have a single cluster that contains all samples.

There are various methods in locating samples in the coordinates, definition or calculation of similarity, particularly in the case of qualitative variables; thus, there are dozens of methods for cluster analysis.

Another, divisive method is often used in machine learning algorithms, such as ID3.

5.3.4 Example of Calculation for Hierarchical Clustering

In this section we explain a popular algorithm.

5.3.4.1 Similarity (Dissimilarity) and Similarity Matrix (Dissimilarity Matrix)

We calculate similarities or nonsimilarities between each of two samples and build them into a similarity or nonsimilarity matrix. Different values are used to represent a similarity for the characteristics of the data vector. Most general values are Euclidean distance, squared Euclidean distance, cosine, Pearson product-moment correlation coefficient, and so on.

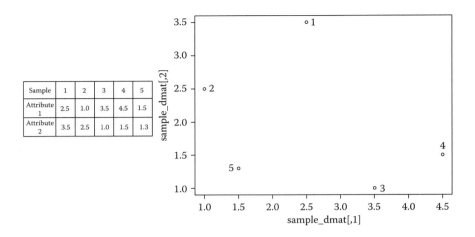

The table shown in the figure:

Sample	1	2	3	4	5
Attribute 1	2.5	1.0	3.5	4.5	1.5
Attribute 2	3.5	2.5	1.0	1.5	1.3

FIGURE 5.16
Example two-dimensional data.

5.3.4.1.1 Similarity and Dissimilarity

Similarity increases as the objects are more similar; oppositely, dissimilarity increases as the objects are more different and decreases as they are more similar. The values such as cosine and correlation efficient are similarity; Euclidean distance and squared Euclidean distance are dissimilarity. In the following, we use an s for similarity and a d for distance or dissimilarity, in short.

5.3.4.1.2 Calculation of Similarity and Dissimilarity

Euclidean distance. A Euclidean distance between two points is derived from their coordinates by applying the Pythagorean theorem. This distance is a dissimilarity measure because it becomes larger as the points move farther away.

The distance between samples 1 and 5 is (5.34) and the one between samples 3 and 4 is (5.35), where $0 \le d_{jk} < \infty$ for samples j and k. Euclidean distances calculated for samples in the example are shown in Table 5.2.

$$d_{15} = \sqrt{(2.5-1.5)^2 + (3.5-1.3)^2} = 2.417$$

$$d_{34} = \sqrt{(3.5-4.5)^2 + (1.0-1.5)^2} = 1.118$$

Squared Euclidean distance. We obtained this before finding the square root in the calculation of Euclidean distance mentioned above. Thus, the squared Euclidean distance between samples 1 and 5 is 5.84. The dissimilarity is enlarged because they remain squared.

TABLE 5.2

Dissimilarity Matrix by Euclid Distance

Samples	1	2	3	4
2	1.803			
3	2.693	2.915		
4	2.828	3.640	1.118	
5	2.417	1.300	2.022	3.007

TABLE 5.3

Similarity Matrix Using Cosines

Samples	1	2	3	4
2	0.971			
3	0.782	0.612		
4	0.809	0.646	0.999	
5	0.972	0.889	0.907	0.924

Cosine. This is a similarity measure in the range of $-1 \leq cos_{jk} \leq 1$ for samples j and k. The closer to 1 the value is, the more similar the samples are. This value is calculated from an angle at the origin, thus the effect of magnification or reduction of scale along the axes. Often in cases of Kansei engineering, it is a suitable characteristic when we obtain similarities by considering whole shapes of sample vectors more important than each evaluation value of a vector. We show a similarity matrix built by cosines in Table 5.3.

$$\cos_{jk} = \frac{(\mathbf{y}_j \cdot \mathbf{y}_k)}{|\mathbf{y}_j||\mathbf{y}_k|} = \frac{\sum_{i=1}^{n} y_{ji} y_{ki}}{\sqrt{\sum_{i=1}^{n} y_{ji}^2} \sqrt{\sum_{i=1}^{n} y_{ki}^2}}$$

5.3.4.2 Clustering Procedure

Next, we perform a procedure of clustering, a reducing process of similarity (dissimilarity) matrix with successive merging samples or clusters.

Step 1a. Merging the most similar (least dissimilar) samples. Merge two samples that have the largest values in similarity matrix (the smallest value in dissimilarity matrix), for example, samples 3 and 4. The cluster made of merged samples 3 and 4 is denoted as (34).

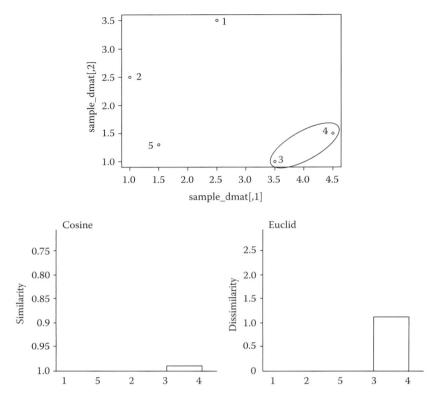

FIGURE 5.17
Created cluster (upper) and dendrograms drawn using cosine (lower left) and Euclidean distance (lower right).

Step 1b. Rebuilding of similarity/dissimilarity matrix. Calculate the similarity between the new cluster (34) and other samples and build a new similarity matrix. There are variations in the clustering method that adopt different representative points of the newly created cluster.

In the methods such as *UPGMA* (unweighted pair-group method using arithmetic averages) and *average linkage between the merged group,* similarity between the cluster created by merging samples (or lower-level clusters) and one of the other samples (or clusters) is calculated by averaging the similarities between each sample in the new cluster and the sample (cluster) in the focus of attention.

We express the procedure mentioned above in the following equations. We obtain a cluster (34); then we calculate the similarities between sample 1 and cluster (34), sample 2 and cluster (34), and sample 5 and cluster (34), respectively. In case of UPGMA, similarity

TABLE 5.4

Reduced Similarity Matrix
Using Cosine

Samples	1	2	5
2	0.971		
5	0.972	0.889	
(3 4)	0.796	0.629	0.916

between the single sample (or each of the samples belonging to the cluster) and the samples belonging to the cluster are averaged and set as a new similarity. Similarities s's are calculated as follows.

In the case of using cosine for similarity (Table 5.4):

$$s_{1(34)} = \frac{1}{2}(s_{13} + s_{14}) = \frac{1}{2}(0.782 + 0.809) = 0.7955$$

$$s_{2(34)} = \frac{1}{2}(s_{23} + s_{24}) = \frac{1}{2}(0.612 + 0.646) = 0.629$$

$$s_{5(34)} = \frac{1}{2}(s_{53} + s_{54}) = \frac{1}{2}(0.907 + 0.924) = 0.9155.$$

Thus, samples merged into a cluster are replaced with the cluster in the similarity matrix, and the reduced matrix is obtained as shown in Tables 5.5 and 5.6.

In the case of using Euclidean distance:

$$d_{1(34)} = \frac{1}{2}(d_{13} + d_{14}) = \frac{1}{2}(2.693 + 2.828) = 2.7605$$

$$d_{2(34)} = \frac{1}{2}(d_{23} + d_{24}) = \frac{1}{2}(2.915 + 3.640) = 3.2775$$

$$d_{5(34)} = \frac{1}{2}(d_{53} + d_{54}) = \frac{1}{2}(2.022 + 3.007) = 2.5145$$

TABLE 5.5

Reduced Dissimilarity Matrix
Using Euclidean Distance

Samples	1	2	5
2	1.803		
5	2.417	1.300	
(3 4)	2.761	3.278	2.515

TABLE 5.6

Reduced (Second) Similarity
Matrix Using Cosine

Samples	2	(1 5)
(1 5)	0.930	
(3 4)	0.629	0.856

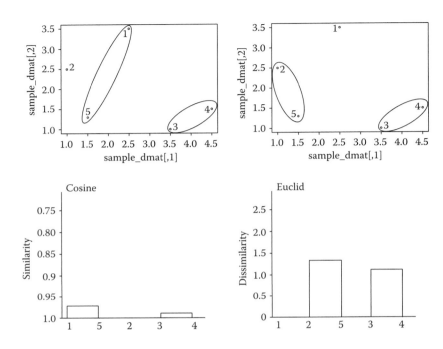

FIGURE 5.18
Created cluster (upper) and dendrograms drawn using cosine (lower left) and Euclidean distance (lower right).

The steps are iterated until we obtain one cluster including all samples.

Step 2a. Merging the most similar (least dissimilar) samples. Next, we merge samples that have the largest similarity (least dissimilarity) in the new reduced matrix. Note that the samples that should be paired are differentiated by (dis)similarity measures. In cosine, most similar samples are 1 and 5. In Euclidian, the closest samples are 2 and 5. This disagreement reflects the difference of the nature of measures, since cosine is a measure related to angles. At the origin (0,0), the angle between samples 1 and 5 is very small, then the most similar pair is judged as 1 and 5. Euclidian measures the distance between the two points. Then, 2 and 5 are the closest. The new similarities between other samples (cluster) and the new cluster are calculated as described below.

Step 2b. Rebuilding of similarity/dissimilarity matrix. Calculate the similarities or dissimilarities between each member of clusters and outer samples or members of the other clusters.

Here, samples 1 and 5 are merged according to the similarity using cosine.

TABLE 5.7

Reduced (Second) Dissimilarity
Matrix Using Euclidian Distances

Samples	1	(2 5)
(2 5)	2.110	
(3 4)	2.761	2.896

$$s_{2(15)} = \frac{1}{2}(s_{21} + s_{25}) = \frac{1}{2}(0.971 + 0.889) = 0.930$$

$$s_{(34)(15)} = \frac{1}{4}(s_{31} + s_{35} + s_{41} + s_{45}) = \frac{1}{4}(0.782 + 0.907 + 0.809 + 0.924) = 0.8555$$

On the other hand, the dissimilarities using Euclidean distances (Table 5.7) between them are as follows:

$$d_{1(25)} = \frac{1}{2}(d_{12} + d_{15}) = \frac{1}{2}(1.803 + 2.417) = 2.110$$

$$d_{(34)(25)} = \frac{1}{4}(d_{32} + d_{35} + d_{42} + d_{45}) = \frac{1}{4}(2.915 + 2.022 + 3.640 \_ 3.007) = 2.896$$

Step 3a. Merging the most similar (least dissimilar) samples. An upper-level cluster is made by merging sample 2 and cluster (1 5) into a new cluster (2 (1 5)), because the similarity between them is the largest in the matrix. With Euclidian dissimilarity, 1 and (2 5) is merged into a new cluster (1 (2 5)).

Step 3b. Rebuilding of similarity/dissimilarity matrix. Then, we calculate the similarity between newly created two clusters:

$$s_{(34)(215)} = \frac{1}{6}(s_{32} + s_{31} + s_{35} + s_{42} + s_{41} + s_{45})$$

$$= \frac{1}{6}(0.612 + 0.782 + 0.907 + 0.646 + 0.809 + 0.924) = 0.780$$

The similarity between clusters (34) and (2(1 5) is an average for six similarities between members of each cluster.

We show Euclidean distances, in addition. The dissimilarily between clusters (34) and (1(2 5) is

$$d_{(34)(125)} = \frac{1}{6}(d_{31} + d_{32} + d_{35} + d_{41} + d_{42} + s_{45})$$

$$= \frac{1}{6}(2.693 + 2.915 + 2.022 + 2.828 + 3.640 + 3.007) = 2.8508$$

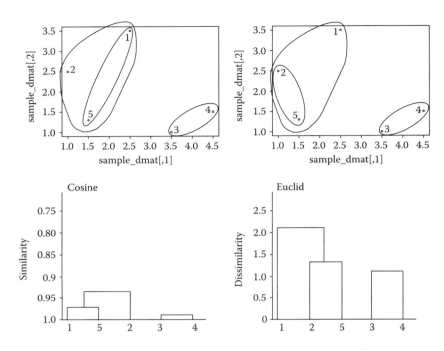

FIGURE 5.19
Created clusters (upper) and dendrograms drawn using cosine (lower left) and Euclidean distance (lower right).

Step 4. Merging the most similar (least dissimilar) samples. Finally, we merge the two clusters into one that contains all samples. The clustering procedure finishes at this step.

The general process of cluster analysis is these steps for iterative merging clusters or samples finally into a single cluster that contains all samples.

5.3.5 Variations of Clustering Methods

Many variations have been developed as clustering methods. Characteristics of some major methods are discussed here.

In the case of UPGMA, as described above, we use averaged similarities or dissimilarities between pairs of all members in each cluster as distances between the clusters.

When we use SLINK (single linkage clustering method), we use the largest similarity (or the smallest dissimilarity) in pairs of members in each cluster. For example, Step 2 for calculating the similarities between each of samples 1, 2, and 5 and cluster (34) are as follows:

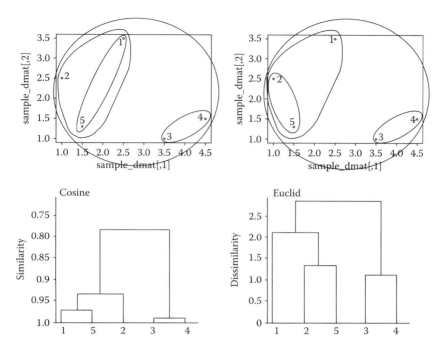

FIGURE 5.20
Created clusters (upper) and final dendrograms drawn using cosine (lower left) and Euclidean distance (lower right).

$$s_{1(34)} = \max (s_{13}, s_{14}) = \max (0.782, 0.809) = 0.809,$$
$$s_{2(34)} = \max (s_{23}, s_{24}) = \max (0.612, 0.646) = 0.646,$$
$$s_{5(34)} = \max (s_{53}, s_{54}) = \max (0.907, 0.924) = 0.924$$

Because the maximum value is used as similarity between clusters in SLINK, similarities between merged clusters typically vary within a narrow range, so that we find branches crowd together in a dendrogram. In some cases, SLINK is preferred, because the final dendrogram is not changed with the order that merges occur, theoretically. However, we do not recommend it in Kansei engineering, because we find a practically innegligible problem that a chain merge of samples often occurs that results in a meaningless categorization.

CLINK (complete linkage clustering method) requires the least similarity or the largest dissimilarity, the opposite of SLINK. The similarities in Step 2 are calculated as follows:

$$s_{1(34)} = \min (s_{13}, s_{14}) = \min (0.782, 0.809) = 0.782,$$
$$s_{2(34)} = \min (s_{23}, s_{24}) = \min (0.612, 0.646) = 0.612,$$
$$s_{5(34)} = \min (s_{53}, s_{54}) = \min (0.907, 0.924) = 0.907$$

The typical dendrogram using CLINK is in a spread shape, also opposite that of SLINK. The other method, UPGMA, has a moderate nature between SLINK and CLINK; thus, it is used most often as a reliable method. In the centroid method, we use distances between the centroid of each cluster as dissimilarity between the clusters. Some researchers advise to use Euclidean distances all the time, and others advise not to be particular about measures; but the result must be easy to understand when Euclidean distances are used.

There are other variations in UPGMA and the centroid method, which uses uneven weight in dissimilarities between clusters. They are WPGMA (weighted pair-group method using arithmetic averages) and W-Centroid (weighted centroid) method. The idea is that when merging clusters A and another one, where cluster A has far more members than others, the similarity between these two clusters is balanced by lighter weighing on similarity between them and cluster A. In case of WPGMA, we weight 1/2 raise to the power of the number of merge the cluster encountered down to the similarity between it and the others. Therefore, similarities between lower-level clusters that are merged in early steps and samples or clusters out of their higher-level clusters are weighted lighter.

5.3.6 Example Analysis of Actual Kansei Evaluation Data

Let us analyze actual data acquired from a Kansei evaluation of many samples. The file beerAveragedLargeSet.txt contains evaluation data of 56 beer cans with 92 Kansei words.

```
> Beer <- read.table("beerAveragedLargeSet.txt,"header=T,
row.names=1)
```

The program used here is written in R and can be installed from the Package Manager. Type the line below, or load the Cluster program from the Packages menu in the menu bar (depending on Windows version).

```
> library(cluster)
```

Here, we use the Agnes program that was named after agglomerative nesting. It was developed by the team led by Professor Peter J. Rousseeuw at Antwerp University in Belgium.

```
> agn1 <- agnes(Beer, metric="euclidian," method="average")
```

There are two options, Euclidean or Manhattan distances, for the metric as a dissimilarity measure. We can choose one of three options, average (between-group average; UPGMA), single (single linkage; SLINK), and complete (complete linkage; CLINK), for method of clustering. Here, we choose Euclidean metric in UPGMA method.

Then, typing the name of variable agn1 shows the result of clustering as we assigned it.

```
> agn1
Call:   agnes(x = Beer, metric = "euclidian", method =
"average")
Agglomerative coefficient:  0.5181686
Order of objects:
  [1] Mack              CASSblue          TexasSelect
LabattBlue        Heineken
  [6] BREWRY            Buckler           LOWENBLAUblue
CASSwhite         DRESSLER
 [11] CoorsLIGHT        MillerLite        SchaeferLIGHT
SAKU              Miller
 [16] NEWQUAY           MALIBU            Budweiser
Carlsberg         CarlsbergSpecial
 [21] PielsRed          Staropramen       Kaisordom
Clausthaller      OldMilwaukee
 [26] YOUNGsLondonLager KOFFred           OldMilwaukeeNA
STROHsNA          STROHsDeep
 [31] MICHELOBgolden    SWANlight         BelgianGold
BRAHMA            CoorsGold
 [36] KARJALA           LOWNBRAUwhite     WestEnd
HITE              OLVI
 [41] KOFFblack         MURPHYs           TIGER
BassPaleAle       WHITBREADpaleAle
 [46] LapinKulta        RoyalDutch        HeinekenDark
TUBORG            BelgiumBrown
 [51] SchulitzBlueOx    COBRA             KARHU
RedBull           PIGsEYE
 [56] REDWOLF
Height (summary):
  Min. 1st Qu.  Median    Mean 3rd Qu.    Max.
  4.047   4.783   5.398   5.731   6.125  10.510

Available components:
[1] "order"      "height"      "ac"        "merge"      "diss"
"call"        "method"      "order.lab"
[9] "data"
```

Agglomerative coefficient (AC) is a measure that indicates the degree that samples are structured, introduced by Kaufman and Rousseeuw (1990). AC is an average of dissimilarities between clusters or samples when merged, where the dissimilarities between samples are standardized within [0, 1]. Kaufman and Rousseeuw suggested that AC will be close to 0 often in the case that samples are almost evenly distributed and there seems lack of explicit cluster structures; however, they did not show any criterion of acceptance.

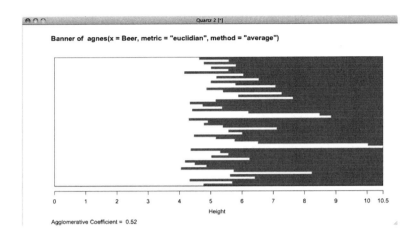

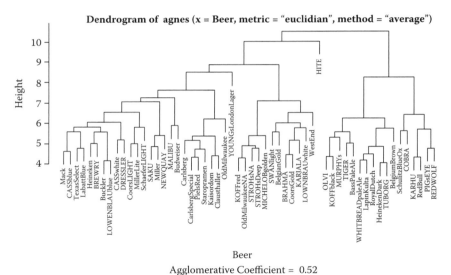

FIGURE 5.21

Banner plot and dendrogram of beer can data.

Typing

```
> plot(agn1)
```

prompts to press Enter. Then, a banner plot and a dendrogram appear. They are shown in Figure 5.21.

The banner plot appears similar as the dendrogram with 90° counterclockwise rotation. It shows the height of each sample when merged; thus, it is useful when we closely find which sample is merged earlier, that is, which sample is closer to others in the cluster.

The problem we encounter when we see a dendrogram is where to cut for finding meaningful clusters. Several methods have been proposed, but on rather weak background.

The common idea in those methods is to find an abrupt change of similarity/dissimilarity. Dissimilarity increases (similarity decreases) monotonically by merging samples or clusters. Abrupt change of dissimilarity (similarity) means that rather different clusters were merged. The number of clusters before the gap should be accounted for with a meaningful number.

The list *height* of data frame agn1 has a history of dissimilarity in the merging process.

```
> agn1$height
 [1]   4.633438   5.585559   4.784415   5.809328   5.015601
5.572568   4.160829   6.046128   5.190135
[10]   6.536083   5.006246   5.798172   7.071142   4.865375
5.398190   7.272158   5.884301   7.632289
[19]   5.162904   4.331931   4.739882   5.369979   4.407026
6.204322   8.482347   8.852662   4.295710
[28]   4.903738   4.781148   5.389744   7.111173   5.570570
6.008410   4.461642   5.166307   5.769535
[37]   6.515402 10.029307 10.510705   4.360691   5.305723
5.569488   5.024938   6.240000   4.177694
[46]   4.486132   4.862643   4.046604   5.740000   8.235587
5.619442   6.410492   4.324711   5.691319
[55]   4.766419
```

The list is ordered by input. It is useful to plot them sorting by their height.

```
> plot(sort(agn1$height), type="b")
```

The type ="b" switch is used for plotting with circles and lines.

The index along the horizontal axis in the plot shows the iteration of merge. The rightmost of the plot is the dissimilarity when all clusters are merged into one. Observing the plot from there to left, we can find a large gap between the 2nd and 3rd points from the right (2 and 3 clusters). Also, there are large gaps between 6th and 7th, and between 9th and 10th points. After that, we see a gentler gradient from the starting point. According to the gaps we mentioned above, we should take 3 clusters, 7 clusters, or 10 clusters for a solution.

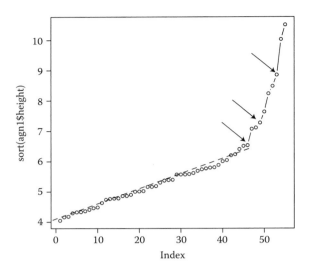

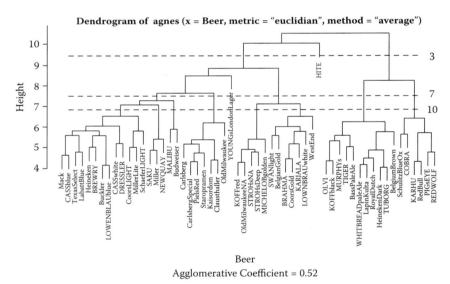

Beer

Agglomerative Coefficient = 0.52

FIGURE 5.22
Dissimilarity plot and dendrogram with cutting points.

5.3.6.1 Finding Features of Clusters

Features of each cluster can be obtained by averaging the evaluation values of the members within the cluster. Here, we first investigate leftmost cluster in the 10-cluster solution.

When we use labeled data, the row can be appointed by typing the label name.

```
> Beer["Mack",]
     High.Grade Ambience Beautiful Unrefined Homely Coarse
Light Smart  Slim
Mack        2.75    3.875    3.875    2.375    2   2.125
3.375 3.125 2.875
     Tender Sweet Stylish Young Simple Soft Elegant Modern
Unique Lite  Warm
Mack       3 2.875        4 3.625    2.5    3        3  3.375
3.5 3.25 1.625
   :            :
```

Set the sum of the evaluation values for members of cluster 1 to data frame "cl1," which contains labels with data.

```
cl1 <- Beer["Mack",]+Beer["CASSblue",]+Beer["TexasSelect",]+Be
er["LabattBlue",]+Beer["Heineken",]+ Beer["BREWRY",]+Beer["Buc
kler",]+Beer["LOWENBLAUblue",]+Beer["CASSwhite",]+ Beer["DRESS
LER",]+Beer["CoorsLIGHT",]+Beer["MillerLite",]+Beer["SchaeferL
IGHT",]
> cl1
     High.Grade Ambience Beautiful Unrefined Homely Coarse
Light Smart  Slim
Mack        36.5    39.75    44.625    33.375   33.5 26.625
47.75 46.25 43.75
   :
```

Convert the data frame "cl1" into a matrix to sort. Here, we convert the data type for sorting the Kansei words by their averaged evaluation value, that is, dividing by the number of members, 13.

```
> cl1mat <- data.matrix(cl1/13)
> cl1mat
     High.Grade Ambience Beautiful Unrefined  Homely   Coarse
Light    Smart
Mack  2.807692 3.057692  3.432692  2.567308 2.576923 2.048077
3.673077 3.557692
:
```

Sort the value for the Mack, the first column, using the Sort command.

```
> sort(cl1mat[1,])
        Coarse       heavytaste    OverDecorated
Gravely              Dark
       2.048077       2.115385       2.125000
2.182692        2.221154
        Heavy          strong        Unseemly
Warm      astringency
       2.336538       2.375000       2.403846
2.432692        2.451923
```

```
        Steady          fullbodied              Sexy
Showy        Original
        2.461538            2.461538          2.509615
2.519231        2.548077
        Unrefined          Homely          Massive
Hard          Grace
        2.567308            2.576923          2.663462
2.682692        2.692308
        Elegant          Gorgeous              Curvy
Soft1        aromatic
        2.701923            2.711538          2.759615
2.788462        2.788462
    Diversified      High.Grade              Chic
Sweet          Polite
        2.798077            2.807692          2.865385
2.913462        2.923077
        bitter            smooth              fruity
Attractive              dry
        2.942308            2.990385          3.000000
3.019231        3.028846
        Basic              Adult          Affected
Ambience          Urban
        3.038462            3.048077          3.048077
3.057692        3.057692
        Feminine            Pretty              Tender
Intelligent          Natural
        3.057692            3.067308          3.076923
3.096154        3.105769
        sweetish        tastesweet          Polished
Soft          Youthful
        3.125000            3.134615          3.144231
3.153846        3.163462
    Fashionable          Modern              Active
mildtaste          Unique
        3.163462            3.182692          3.201923
3.211538        3.221154
    Individual      Monotonous          Healthy
Refined    Nice.looking
        3.240385            3.240385          3.288462
3.298077        3.298077
        Childish            Calm              Lively
Masculine          Slim
        3.317308            3.317308          3.336538
3.336538        3.365385
        soursweet        Beautiful              Formal
Sporty        Natural2
        3.365385            3.432692          3.432692
3.480769        3.490385
```

	Sharp	Smart	Straight
Casual		Bright	
	3.509615	3.557692	3.567308
3.586538		3.605769	
	tastesour	cleanly	Stylish
Light		cool2	
	3.615385	3.653846	3.663462
3.673077		3.692308	
	Simple	drinkable	Young
Fresh		Refreshing	
	3.721154	3.778846	3.836538
3.836538		3.865385	
	Plain	Cool	Lite
refreshingtaste		clearcut	
	3.932692	3.932692	3.942308
4.048077		4.086538	
	crisp	lighttaste	
	4.125000	4.163462	

We can find that *light taste*, *crisp*, *clear-cut* and *refreshing taste* obtained large evaluation values. Most of the members of this cluster are painted in cold colors, such as white, blue, green, or silver. The rest of them are in achromatic color, that is, white or shiny metallic. Therefore, we find the correspondence between Kansei and color.

For the Cluster 10, perform the analysis in the same way.

```
> cl10 <- (Beer["SchulitzBlueOx",]+Beer["COBRA",]+-
Beer["KARHU",]+Beer["RedBull",]+
+ Beer["PIGsEYE",]+Beer["REDWOLF",])
> cl10mat <- data.matrix(cl10/6)
> sort(cl10mat[1,])
```

	Tender	Sweet	Feminine
Pretty		Soft	
	1.354167	1.395833	1.416667
1.520833		1.645833	
	Soft1	fruity	Light
Slim		Polite	
	1.687500	1.750000	1.854167
1.937500		1.979167	
	Refreshing	Smart	Elegant
tastesweet		Natural	
	2.000000	2.041667	2.083333
2.083333		2.104167	
	Cool	Lite	Simple
Fresh		lighttaste	
	2.145833	2.166667	2.187500
2.187500		2.187500	

```
      Childish            Sexy       Gorgeous
Bright         soursweet
        2.208333          2.208333       2.250000
2.312500          2.312500
        Homely          sweetish          Warm
refreshingtaste         Healthy
        2.354167          2.354167       2.375000
2.395833          2.416667
        Plain            Casual       Monotonous
Calm           Urban
        2.437500          2.437500       2.437500
2.458333          2.458333
        Modern             Grace       Beautiful
Formal         drinkable
        2.479167          2.479167       2.520833
2.541667          2.541667
        Youthful          Unseemly    High.Grade
Fashionable         Sporty
        2.562500          2.583333       2.625000
2.666667          2.687500
      Attractive          Ambience    Intelligent
Natural2          Basic
        2.708333          2.729167       2.770833
2.770833          2.854167
        Curvy             smooth          Young
Straight         Polished
        2.916667          2.916667       2.937500
2.958333          2.958333
      Unrefined            cool2        Refined
clearcut          Affected
        3.000000          3.000000       3.020833
3.020833          3.041667
        crisp            mildtaste       Coarse
Stylish            Chic
        3.041667          3.125000       3.187500
3.208333          3.229167
        Showy      OverDecorated       aromatic
Nice.looking           Lively
        3.270833          3.270833       3.270833
3.354167          3.416667
        Unique             Sharp      Diversified
Dark            Adult
        3.562500          3.562500       3.583333
3.625000          3.666667
      tastesour           Massive       Original
Active           Steady
        3.729167          3.750000       3.854167
3.895833          3.958333
```

FIGURE 5.23
Cans belong to Cluster 1 (upper) and Cluster 10 (lower).

```
      astringency            cleanly            Gravely
dry        heavytaste
         4.000000           4.083333           4.229167
4.229167            4.229167
           Heavy              bitter          fullbodied
strong        Individual
         4.250000           4.333333           4.375000
4.395833            4.416667
           Hard             Masculine
         4.437500           4.645833
```

Highly evaluated words are *masculine, hard, individual, strong,* and *full-bodied.* These cans have animals or one-eyed pirates on them.

The example where we find decisive design elements from the results of cluster analysis is described in Section 6.3.

5.4 Linear Regression Analysis

The section explains linear regression analysis. This is related to the basis of quantification theory Type I, which has been intensively used in Kansei/affective engineering studies.

Suppose that we approximate or explain the value of a variable y using p variables $x_1, x_2, \ldots, x_p$. We can obtain this sort of a mathematical model from multiple linear regression analysis. In this case, $x_1, x_2, \ldots, x_p$ are called explanatory variables (also called independent variables), and y is called an objective variable (also called a dependent variable). However, variables x_1, x_2, ... , x_p are assumed to be independent of one another.

In the case of two explanatory variables, that is, $p = 2$, a multiple linear regression model is as follows:

$$y_i = a_1 x_{1i} + a_2 x_{2i} + b$$

where the variable y is for the ith sample. Generally, the task of multiple regression analysis is to solve a_1 and a_2, the weights for variables x_1 and x_2, respectively, using all samples.

If the weight a_1 is positive and a_2 is negative, larger x_1 makes y larger, and larger x_2 makes y smaller. In case that x_1 affects y stronger than x_2, the absolute value of a_1 is larger than that of a_2. When we solve the weights, as mentioned above, we can understand that each explanatory variable has a positive or negative effect, or which explanatory variables affect the objective variable more strongly.

5.4.1 Mathematical Solution of Simple Regression

At first, we will follow simple regression ideas. The simple regression model has only one explanatory variable (x). Then, the simple regression model is expressed as

$$\hat{y} = ax + b$$

$\hat{y}$ (y hat) means the estimated value of y.

1. To ease solving, let x and y convert to deviation data from the mean. This procedure is often called centering. $\bar{x}$ is the mean of x and $\bar{y}$ is the mean of y, those have m samples. i means sample i.

$$x_i = x_i - \bar{x}, \quad y_i = y_i - \bar{y}$$

2. By subtracting each mean of x and y, the regression line goes through the origin. It is removing the bias b. Then, cumulative errors between measured y and regression equation ax_i are defined as

$$F = \sum_{i=1}^{n} \left(y_i - \hat{y} \right)^2 = \sum_{i=1}^{n} \left(y_i - ax_i \right)^2$$

Then, regression is turned into a minimizing problem of F.

3. Expansion is as follows:

$$\sum (y_i - ax_i)^2 = \sum y_i^2 - 2a \sum x_i y_i + a^2 \sum x_i^2$$

4. Each element of #3 is expressed in vector forms. Superscript T means transpose.

$$\sum y_i^2 = \mathbf{y}^T\mathbf{y}, \quad \sum x_i y_i = \mathbf{x}^T\mathbf{y}, \quad \sum x_i^2 = \mathbf{x}^T\mathbf{x}$$

$$\sum y_i^2 - 2a \sum x_i y_i + a^2 \sum x_i^2 = \mathbf{y}^T\mathbf{y} - 2a\mathbf{x}^T\mathbf{y}, + a^2\mathbf{x}^T\mathbf{x}$$

5. Since the problem is a minimizing problem for deriving the slope of *a*, partial differentiation on *a* is performed.

$$\frac{\partial F}{\partial a} = -2\mathbf{x}^T\mathbf{y} + 2a\mathbf{x}^T\mathbf{x} = 0$$

$$\therefore 2(a\mathbf{x}^T\mathbf{x} - \mathbf{x}^T\mathbf{y}) = 0$$

6. Solving *a*. Multiplying with 1/*number_of_samples* makes the denominator the covariance of *x* and *y* (*Sxy*), and the numerator the covariance of *x* and *x* (*Sxx*). Note that denominator is *n*, not (*n*–1).

$$a = \frac{\mathbf{x}^T\mathbf{y}}{\mathbf{x}^T\mathbf{x}} = \frac{\frac{1}{n}\mathbf{x}^T\mathbf{y}}{\frac{1}{n}\mathbf{x}^T\mathbf{x}} = \frac{Sxy}{Sxx}$$

Covariance of two vectors is (1/elements number) × sum of elements.

7. Bias *b* is obtained by this relationship:

$$\bar{y} = a\bar{x} + b \text{ then, } b = \bar{y} - a\bar{x}$$

5.4.2 Solving a Simple Regression Example with R

An example of Kansei evaluation data is a ratio variation of white boxes (Ishihara et al., 2001a; 2001b; 2003). Nine boxes that have a different right side

TABLE 5.8

Proportion (Length of a Side)
and Kansei Evaluation on *calm*

Proportion	*Calm* Evaluation
Xi	Yi
1.000	3.889
1.200	3.605
1.414	3.531
1.620	3.469
1.732	3.568
2.000	3.506
2.236	3.235
2.645	3.086
3.000	3.198

length were used for evaluation stimuli. Ratios of the side are 1:1,1.2, 1.414 ($\sqrt{2}$), 1.62(golden section), 1.732 ($\sqrt{3}$), 2.0, 2.236 ($\sqrt{5}$), 2.645($\sqrt{7}$), 3.0. Stimuli were drawn by 3D CG and were made into a leaflet. The forehand side length of the box is 11 cm.

Thirty-two pairs of Kansei words were used for the questionnaire. Eighty-one subjects (53 female, 28 male, ages 18–25) participated in the experiment.

Regression analysis was used to reveal relationships between side length and Kansei evaluation. Side length was assigned as an explanatory variable (a regressor) [x], and evaluation value (averaged between subjects) on a Kansei word was assigned as an objective variable [y].

We show the calculation example of Kansei word *calm* and box proportion (Table 5.8). Figure 5.24 shows that the longer box is less *calm* and the cube box is the most *calm*. The exact relation was analyzed using the R commands shown below. lm means linear model. The symbol ~ is commonly used in many aspects of R language. A ~ B means that A is explained or predicted by B.

```
> proportion <- c(1.0,1.2,1.414,1.62,1.732,2.0,2.236,2.645,3.0)
> calm <- c(3.889,3.605,3.531,3.469,3.568,3.506,3.235,3.086,3.
198)
> plot(calm ~ proportion, ylim=c(2,4))
> abline(lm(calm ~ proportion))
> lm(calm ~ proportion)

Call:
lm(formula = calm ~ proportion)

Coefficients:
(Intercept)    proportion
    4.0821       -0.3355
```

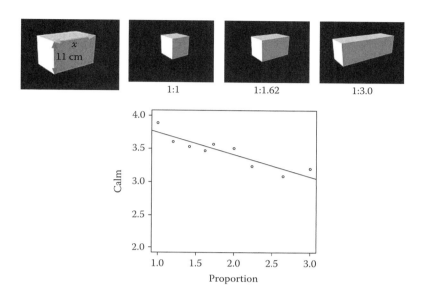

FIGURE 5.24
Plot of *calm* on box ratio evaluation and regression line.

The obtained regression model is $y = -0.3355x + 4.0821$. As x gets larger, evaluation of y decreases. When the box ratio increases to 1.0, then evaluation of *calm* will decrease to 0.3355.

Verification of the solution, by calculation of (Sxy/Sxx), is shown below.

```
> t(c(proportion - mean(proportion))) %*% c(calm-mean(calm))
          [,1]
[1,] -1.18209
> t(c(proportion - mean(proportion))) %*% c(proportion -
mean(proportion))
          [,1]
[1,] 3.523629
> -1.18209 / 3.523629
[1] -0.3354752
> mean(calm) - (-0.3354752)* mean(proportion)
[1] 4.082083
```

5.4.3 Mathematical Solution of Multiple Regression

When the regression model has plural explanatory variables, the model is called multiple regression. The next expression shows a multiple regression model consisting of two independent variables. x_1 has weight a_1 and x_2 has weight a_2.

$$\hat{y} = a_1 x_1 + a_2 x_2 + b$$

1. Each variable is centered, just as in the simple regression model procedure.

$$x_{1i} = x_{1i} - \overline{x}_1, \quad x_{2i} = x_{2i} - \overline{x}_2, \quad y_i = y_i - \overline{y}_i$$

2. Then, cumulative errors between measured y and regression equation are determined as the following equation. To solve and view it with ease, $(y_i - \hat{y})^2$, which has the same meaning as $(\hat{y} - y_i)^2$, was changed.

$$F = \sum_{i=1}^{n} \left(\hat{y} - y_i \right)^2 = \sum_{i=1}^{n} \left(a_1 x_{1i} + a_2 x_{2i} - y_i \right)^2$$

3. To expand it, let $a_1 x_{1i} + a_2 x_{2i}$ as Δ, y_i as Ξ. Then, the equation can be expressed as

$$\sum_{i=1}^{n} \left(\Delta - \Xi \right)^2 = \sum_{i=1}^{n} \left(\Delta^2 - 2\Delta\Xi + \Xi^2 \right)$$

Using this well-known expansion, we obtain

$$\sum_{i=1}^{n} \left(a x_{1i} + a x_{2i} - y_i \right)^2 =$$

$$\sum_{i=1}^{n} \left(a_1^2 x_{1i}^2 + 2a_1 \, x_{1i} \, a_2 x_{2i} + a_2^2 x_{2i}^2 - 2a_1 \, x_{1i} y_i - 2a_2 \, x_{2i} y_i + y_i^2 \right)$$

Using vector expression, F can be written as

$$F = a_1^2 \mathbf{x}_1^T \mathbf{x}_1 + 2a_1 \, a_2 \mathbf{x}_1^T \mathbf{x}_2 + a_2^2 \mathbf{x}_2^T \mathbf{x}_2 - 2a_1 \, \mathbf{x}_1^T \mathbf{y} - 2a_2 \, \mathbf{x}_2^T \mathbf{y} + \mathbf{y}^T \mathbf{y}$$

4. The problem to solve is a minimizing problem for deriving the slope of a_1 and a_2.

$$\frac{\partial F}{\partial a_1} = 2a_1 \, \mathbf{x}_1^T \mathbf{x}_1 + 2a_2 \mathbf{x}_1^T \mathbf{x}_2 - 2\mathbf{x}_1^T \mathbf{y} = 0$$

$$\frac{\partial F}{\partial a_2} = 2a_1 \, \mathbf{x}_1^T \mathbf{x}_2 + 2a_2 \mathbf{x}_2^T \mathbf{x}_2 - 2\mathbf{x}_2^T \mathbf{y} = 0$$

5. Solving a1 and a2,

$$\begin{cases} S_{11}a_1 + S_{12}a_2 = S_{1y} \\ S_{12}a_1 + S_{22}a_2 = S_{2y} \end{cases}$$

We can calculate S_{11}, S_{22}, S_{12}, S_{1y}, S_{2y} from centered data. a_1 and a_2 are called partial correlation coefficients.

6. Bias b is obtained by this relation. Substitute a_1 and a_2 by obtained weight, and substitute $\bar{y}$, $\bar{x}_1$, $\bar{x}_2$ as calculated value.

$$\hat{y} - \bar{y} = a_1(x_1 - \bar{x}_1) + a_2(x_2 - \bar{x}_2)$$

5.5 Quantification Theory Type I

5.5.1 Purpose of Quantification Theory Type I in Kansei/Affective Engineering

Quantification theory Type I (QT1) most often has been used to analyze direct and quantitative relationships between a Kansei word and design elements. For instance, you may want to understand whether a *feminine* design is based on a certain color or illustration, or which is more important as a design element. In another case, you may want to know which color makes the product attractive or unattractive, and how to combine colors quantitatively to make it more attractive. QT1 is an effective analysis method for building a mathematical model of the relationships between a Kansei word y and two or more design elements x_1, x_2, x_3.... The results obtained from QT1 can be stored in a Kansei database or transformed into a knowledge base and integrated into a Kansei/affective engineering expert system.

5.5.2 Concept of QT1

Multiple linear regression analysis deals with intervals or proportional scaled explanatory variables. However, design elements such as color selection and the presence or absence of functions or illustrations cannot be expressed by quantity or order (i.e., they are nominal scaled variables).

QT1 is an expansion of the multiple linear regression analysis method. It deals with nominal scaled explanatory variables and an interval or a proportionally scaled objective variable. In short, we can assign each variation in a design element to a nominal scaled explanatory variable and then perform multiple linear regression analysis. In the quantification theory, a design element is referred to as an *item*, and each variation in a design element is a *category*.

For example, when we have 10 different colors of beer cans (i.e., 10 categories), we prepare 10 explanatory variables and assign a category to each variable. To express that the color is silver, we assign a value of 1 to an explanatory variable that corresponds to silver and assign 0 to all the rest of the explanatory variables for item color. Such an explanatory variable that is assigned either a 1 or a 0 to express the presence or absence of a design variation as a nominal scaled variable is often called a dummy variable.

When we analyze Kansei evaluation data, we can set an evaluation value for a Kansei word to an objective variable y. An averaged evaluation value among participants is often used for such an objective variable. Additionally, we divide our design elements into items and their categories, and then set each category of an item to 1 or 0 for dummy variable x's to use as explanatory variables. An item of a product is a design element (variable) that is supposed to contribute to Kansei. This could be a color, a shape, or the location of a logo, for example. A category is a specification of an item. For instance, categories of an item color may be white, black, or red, and so on. Categories of an item location could be top, center, or bottom.

Assume an objective variable for an evaluation with a Kansei word y_λ for each sample λ ($\lambda = 1, ..., m$). An objective variable y_λ is linearly related to explanatory variables, in which each dummy variable x expresses the presence or absence of a design item and categories. When the set of samples has two items with two and three categories, we represent the relationship between an objective variable and its explanatory variables with

$$y_\lambda = a_{11}x_{11\lambda} + a_{12}\, x_{12\lambda} + a_{21}\, x_{21\lambda} + a_{22}\, x_{22\lambda} + a_{23}\, x_{23\lambda} + \varepsilon_\lambda,$$

$$\lambda = 1, 2, ..., m$$

where $x_{ij\lambda}$ is a dummy variable that expresses the presence or absence of the item i with the category j for sample λ. Only a variable assigned as a category of an item is set to 1, and the variables corresponding to the all the rest of the categories of the same item are set to 0. A coefficient of a dummy variable for item i with the category j is denoted as a_{ij}. We refer to this as the category score or category weight. Additionally, m is the number of samples and ε_λ is a residual.

Then, let us estimate a vector of category scores A, given data vectors Y and X. We can follow a similar process used in multiple linear regression analysis. We will obtain $\hat{a}^{ij}$, the estimate of a_{ij}, where the sum of squares of residuals $\varepsilon_\lambda = y_\lambda - \hat{y}_\lambda$ is minimum. Suppose that the number of samples m is sufficiently large to solve these equations. A number of samples greater than twice the number of dummy variables is recommended. A category score expresses the degree of contribution and the direction of its category for a specific Kansei. Additionally, we can build a model to estimate the Kansei evaluation $\hat{y}_\lambda$ using the estimated category scores and dummy variables. We can use this model to estimate Kansei responses to new design products.

The correlation matrix between each design item and its Kansei evaluation leads to a partial correlation coefficient ρ between a Kansei y and each design item. This indicates the degree to which item i affects Kansei y.

The multiple correlation coefficient R determines how well the estimation model fits the observed data (i.e., the precision of the estimation model). In our experience, R *should* be more than 0.8 and *must* be more than 0.6 for a Kansei evaluation data model.

5.5.3 Analysis of Beer Can Design Using QT1

We next present an example analyzing the evaluation data of beer cans. The cans are the same 56 used in Section 5.1. There are a great many design elements even in a cylindrical can. We selected more than 20 items from this set of cans and obtained more than 70 categories in total. However, in the case where the number of explanatory variables is greater than the number of samples (i.e., greater than the number of objective variables), we cannot obtain definite solutions from the simultaneous equations built on multiple regression analysis. This is the limitation of the least squares method used in multiple regression analysis. In such cases, partial least squares should be used. This is explained in Section 5.6.

Therefore, an effective strategy would be to select predominant design items first by performing the cluster analysis described in Section 5.3, and then to analyze the relationships between the predominant design items and Kansei. In this case, we used colors, illustrations, and shapes of labels in the following analysis, because they are closely related to Kansei. Table 5.9 shows these categories and design items. There were 10 categories for item

TABLE 5.9

Numeric Codings of Categories in Beer Can Design Items

Can Color	Can Illustration	Label Shape
1: White	1: Animal	1: Oval Shape
2: Silver	2: Bird	2: Other Trad
3: Gold	3: Person	3: None of Above
4: Blue	4: Barley/Hops	
5: Black	5: Crown/Symbol	
6: Red	6: Other Object	
7: Green	7: No Illustration	
8: Cream		
9: Light Blue		
10: Yellow		

TABLE 5.10

Evaluation Values for *Bitter* on 56 Beer Cans

	Name of Beer	Colors	Illustration	Label	Bitter
1	Mack	4	1	1	3.25
2	Koff Red	6	5	3	3.63
3	Olvi	5	6	3	4.50
4	Coors Light	2	5	3	3.13
5	Stroh's NA	3	5	3	4.00
6	Carlsberg	1	5	3	2.38
7	Texas Select	1	4	3	3.00
8	Cass White	1	5	3	2.25
9	Schlitz Blue Ox	2	1	3	4.13
10	Saku	4	5	1	3.38
11	Murphy's	8	5	2	4.13
12	Tiger	4	5	1	3.75
13	Newquay	4	6	2	3.50
14	Miller	1	2	3	3.00
15	Swan Light	1	2	1	3.25
16	Carlsberg Special	1	6	3	2.50
17	Belgian Gold	3	4	3	2.50
18	Bass Pale Ale	5	5	1	4.38
19	Karhu	5	1	3	3.88
20	Brahma	3	4	3	3.75
21	Michelob Golden	3	5	1	4.13
22	Staropramen	8	4	3	2.00
23	Lowenbrau White	8	5	3	2.38
24	Piel's Red	1	5	3	2.25
25	Malibu	2	6	3	3.13
26	Whitbread Pale Ale	4	5	1	4.00
27	Cass Blue	4	7	3	3.38
28	Kaisordom	1	5	1	2.88
29	Pig's Eye	3	3	3	4.38
30	Heineken	7	5	1	3.63
31	Cobra	5	1	3	4.50
32	Dressler	2	5	3	2.63
33	Redwolf	6	1	1	4.38
34	Old Milwaukee	6	5	3	2.75
35	Labatt Blue	4	7	3	2.38
36	West End	3	7	3	3.63
37	Clausthaller	1	5	3	2.63
38	Buckler	1	5	1	2.50
39	Young's London Lager	10	1	3	2.00
40	Royal Dutch	7	5	1	3.13

TABLE 5.10 (continued)

Evaluation Values for *Bitter* on 56 Beer Cans

	Name of Beer	Colors	Illustration	Label	Bitter
41	Red Bull	5	1	3	4.75
42	Lapin Kulta	4	7	3	3.25
43	Karjala	6	5	1	3.50
44	Belgium Brown	6	4	3	4.00
45	Old Milwaukee NA	6	5	3	3.25
46	Heineken Dark	6	5	1	4.25
47	Tuborg	7	5	1	4.25
48	Hite	8	7	3	2.88
49	Budweiser	1	5	3	2.88
50	Schaefer Light	2	4	1	3.63
51	Brewry	1	5	3	3.00
52	Koff Black	5	5	3	4.38
53	Lowenbrau Blue	9	5	3	2.25
54	Miller Lite	6	2	3	3.25
55	Stroh's Deep	6	5	3	3.75
56	Coors Gold	3	5	3	3.13

color; 7 categories for item illustration (animal, bird, person, barley or hops, crown or symbol of nobility, other object, no illustration), and 3 categories for item label shape (oval, some other traditional shape, some different shape other than the first two, no label). Here, we used the evaluation value for *bitter* (Table 5.10) as an objective Kansei word averaged over all participants.

We used a QT1 program written in R (http://aoki2.si.gunma-u.ac.jp/R/qt1.html) by Professor Shigenobu Aoki, who teaches at the Faculty of Social and Information Studies at Gunma University. Simply typing qt1(*data*)causes his program to perform all calculations for QT1.

We input the data in row order from Table 5.10 and assigned the parameter ncol=4 to specify that the input data is folded with four columns in a line. We can adjust this parameter to fit the number of items we use.

```
> designElements <- matrix(c(
4,1,1,
6,5,3,
  :
6,5,3,
3,5,3
), ncol=3, byrow=TRUE)
```

Last row of the data does not need "," at its line end

```
> designElements <- data.frame(designElements)
> designElements[,1:3] <- lapply(designElements, factor)
```

```
> y <- c(3.25, 3.63, 4.50, 3.13, 4.00, 2.38, 3.00, 2.25, 4.13,
3.38, 4.13, 3.75, 3.50, 3.00, 3.25, 2.50, 2.50, 4.38, 3.88,
3.75, 4.13, 2.00, 2.38, 2.25, 3.13, 4.00, 3.38, 2.88, 4.38,
3.63, 4.50, 2.63, 4.38, 2.75, 2.38, 3.63, 2.63, 2.50, 2.00,
3.13, 4.75, 3.25, 3.50, 4.00, 3.25, 4.25, 4.25, 2.88, 2.88,
3.63, 3.00, 4.38, 2.25, 3.25, 3.75, 3.13)
```

We can see all results using the qt1 function as follows:

```
> summary(qt1result)
$coefficients
        Category Scores
X1.1          -0.519752
X1.2           0.098237
X1.3           0.325144
X1.4          -0.218241
X1.5           1.078849
X1.6           0.336625
X1.7           0.097409
X1.8          -0.579906
X1.9          -0.845717
X1.10         -1.413847
X2.1           0.245550
X2.2           0.073705
X2.3           0.886558
X2.4          -0.098692
X2.5          -0.072580
X2.6          -0.161242
X2.7           0.117600
X3.1           0.299100
X3.2           0.984913
X3.3          -0.177774
bias           3.346071

$partial
   partial correlation coefficient    t-value        P-value
X1     0.81681 10.2097 5.0513e-14
X2     0.40112  3.1576 2.6475e-03
X3     0.59490  5.3370 2.0967e-06

$prediction
    Observed  Predicted  residual
#1     3.25 3.6725 -0.4224803
#2     3.63 3.4323  0.1976583
       :       :       :
#56    3.13 3.4209 -0.2908616

attr(,"class")
[1] "qt1"
```

Partial correlation coefficients of the items are shown in $partial. The maximum is the partial coefficient of Item 1, color; that is, color is most closely related to *bitter*. The next is the shape of the label (Item 3), and the least is the illustrations (Item 2).

Category scores, which express the degree of relevance with Kansei evaluation, are shown as $coefficient. We can draw them in a bar graph using the plot command.

```
> plot(qt1result)
```

The bars of 1.1 (Item 1, Category 1), 1.2, ..., 2.1, ..., 3.3 are stacked from the bottom to the top of the graph (Figure 5.25).

Because we have 10 colors, the bars from the 1st bar at the bottom to the 10th bar show their category scores. The 5th category, black, has the largest positive score. In contrast, the 10th category, yellow, has the largest negative score. Thus, the beer seems most *bitter* for black and least *bitter* for yellow.

We have seven kinds of illustrations. The category that has the largest positive category is the third one, a person's face. The largest negative category is given to other symbols. Thus, the illustration of a person's face indicates a beer that is the most *bitter*. The evaluation of Pig's Eye, with a pirate's face,

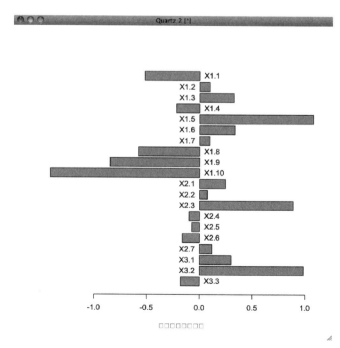

FIGURE 5.25
Graph of category scores.

is considered to be the source of this. The next largest score is for category 1, illustrations of an animal.

As for label shape, category 2 (other traditional shape) has the largest positive score, that is, a label of some nontraditional shape goes with beer that is most *bitter*.

In general, we can obtain clear rules that are easy to understand and agree with common intuition from the results derived from QT1. Additionally, we often obtain interesting rules in relationships between Kansei and design elements that are unexpected even by designers.

QT1 is an effective tool for the analysis of Kansei, especially for products where the relationships between design and Kansei are not clearly understood. The results of analysis using QT1 on customers and designers have created a considerable amount of Kansei design knowledge and many successful products.

At the same time, the model of QT1 assumed linear relationships between Kansei and the design elements. This is a weak point in that we cannot directly incorporate combination effects of design elements or nonlinear relationships between the design and Kansei into the QT1 model. A better approach would be to use an analysis method that incorporates a nonlinear model, as explained in later sections.

NOTE: We thank Professor Shigenobu Aoki, who gave permission for the use of his R program.

5.6 Partial Least Squares Regression

Partial least squares (PLS) was developed by Swedish econometrician Herman Wold and co-workers in the mid-1970s. The most popular applications of PLS have been in the chemometrics field since the mid-1990s. A typical example takes spectrum distribution on a huge number of x. In these applications, the numbers of x up to several hundreds and correlations between x variables are very high because of the spectrum. On the other hand, y takes a measured value such as temperature or PH, and the sample number is 10s at the most. Brereton (2003) shows such smaller sample size cases in chemometrics. Common multiple regression cannot deal with such data.

In Kansei/affective engineering, the relationship between Kansei word evaluation and design elements had been analyzed with QT1. QT1 is a deterministic method because it is a variation of the multiple regression model and its solving method uses the least squares method. Although QT1 is widely used, there are two defects. The first is a problem of sample size, and QT1 incorporates dummy variables. In a multiple regression model, simultaneous equations could not have been solved when the number of variables exceeds the number of samples. In Kansei/affective engineering, many cases have a larger number of design variables than of samples. Then, the analyst has to divide design variables to do analysis. The second defect is a problem of interactions between x variables; if there are heavy interactions between x variables, its analyzing result is distorted. This problem is known as multicollinearity in multiple regression analysis. PLS has the possibility to resolve both problems.

5.6.1 PLS Structure and Algorithm

PLS uses several latent variables. There are s (number of samples) observations of objective (dependent) variable. These become vector $\mathbf{y}$. There are p dimensional explanatory (independent) variables. These become vector $\mathbf{x}$. There are number s of $\mathbf{x}$. They become matrix $\mathbf{X}$. The algorithm given below is based on Miyashita and Sasaki (1995).

At the first step, $\mathbf{w}$, the covariance vector of $\mathbf{y}$ and $\mathbf{x}$ is computed. The $\mathbf{w}$ is treated like an eigenvector in PCA. Second, the latent variable t_1 is introduced. Output from t_1 ($t_1 = \Sigma x_{ik}w_k$, thus $\mathbf{t}_1 = \mathbf{Xw}_1$) is regarded as principal component score. Third, l_{11}, l_{12}, correlations between $\mathbf{x}$ and t_1 (these compose vector $\mathbf{l}_1$) are computed. They correspond to principal component loadings (correlation between principal component scores on a principal component and an original variable). Fourth, q_1, the relation between $\mathbf{t}_1$ and $\mathbf{y}$, is computed. The q_1 is the result of single regression analysis (with no bias term), which takes $\mathbf{t}_1$ as an explanation variable and $\mathbf{y}$ takes as objective variable. Fifth, x–t_1–y relation can be computed. Sixth, second latent variable t_2 is introduced and we compute x–t_2–y relation with the same procedure noted above. This time, y takes the residual of x–t_1–y model, and $\mathbf{X}$ takes X residual of x–t_1–y model, which was obtained by estimation by inverse way ($\mathbf{X}$new $= \mathbf{X} - t_1\mathbf{l}_1^\mathsf{T}$). As the result, relations between two latent variables and y or x are obtained. Finally, we get a regression equation by composing these relations.

The high-dimensional $\mathbf{x}$ is projected onto a smaller dimensional orthogonal space. An example is shown in the analysis section of Chapter 11 of this book. The relation between the projection and y is solved with simple regression. Thus, the dimensionality problem (sample size problem) is solved. The projection procedure is similar to the procedure of PCA. Since the projection is a linear transformation, regression coefficients can be computed. Thus, correlations between explanation variables do not cause the multicollinearity problem. Multicollinearity is also avoided since there is no need to solve simultaneous equations.

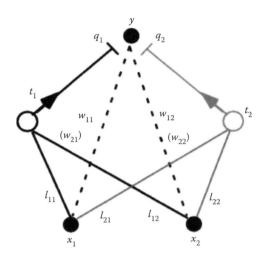

FIGURE 5.26
Structure of PLS (single y variable).

5.6.2 PLS Analysis of Personal Garden Kansei Evaluation Data and Comparison with QT1

The mathematical features of PLS are quite attractive, but there is no statistical pointer for acceptance of the number of x variables. Since this study is the first attempt to use PLS in Kansei/affective engineering, we should consider its ability. We shot more than 150 photos of residential gardens at Nagano, Hokkaido, and Hiroshima, and 47 panoramic photos were chosen for evaluation. These photos were projected on the screen by an LCD projector. The subjects were 19 unpaid university students (13 male and 6 female). The SD questionnaire contained 26 Kansei word pairs such as *calm in mind* [ ] [ ] [ ] [ ] [ ] *not calm in mind*. The evaluation was done in 2004. We compared analyzing results of PLS and QT1 with Kansei evaluation data on these 47 residential gardens. The design element table has 32 items and 89 categories. PLS implementation, which we used, was JMP 5.2 (SAS).

To analyze the data by QT1, we divided design elements into five (23/9/18/28/11) categories. We performed QT1 on each division, and we got five results. We compared multiple correlation coefficients (correlation between predicted y values and measured values) of QT1 and PLS. Even when incorporating 89 variables, PLS's multiple correlation coefficient was much higher than QT1. Thus, PLS makes a model that fits the data better than QT1.

We obtained a numerically excellent result with PLS. The analyzing result seems a nearly perfect fit to the data. Another side of the perfect fit is the overfit to the data. Overfitting to the data is picking up all of the (unwanted) deviation of the sample, and it makes the model too complex. As a result,

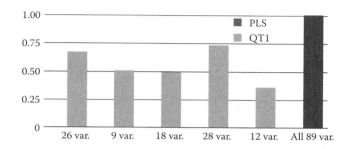

FIGURE 5.27
Comparison between PLS and QT1 results (multiple correlation coefficients).

an overfitted model becomes more specific than generalized. We also compared the QT1 result with the PLS result in a qualitative manner. Although PLS was analyzed with all 89 x variables as we have shown above, in this comparison, we compute the model by adding variables in five steps. Step 1 used 22 variables, step 2 used 22 + 9 = 31 variables, step 3 used 31 + 18 = 49, step 4 used 49 + 28 = 77, and finally step 5 used 77 + 12 = 89 variables. The numbers of latent variables used were four at steps 1 and 2, and six at steps 3, 4, and 5. These numbers were decided to have best performance (smaller residuals) by several trials with different numbers of latent variables. Next we surveyed accordance of ranks of categories in each item, between PLS and QT1. Figure 5.27 shows the accordance along the number of variables.

When the number of x variables exceeds the sample size, the accordance slightly decreases (right side of Figure 5.28). It seems that in the cases of

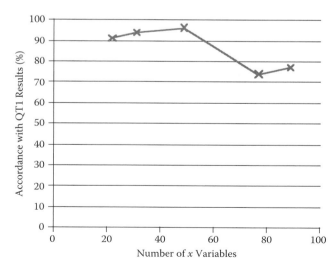

FIGURE 5.28
Accordance with PLS and QT1 results.

smaller sample size, averaging effect is less than smaller deviations reflecting the result. In many practical cases of Kansei engineering, we can get the numerically accurate model of relations between design elements and Kansei with PLS. In other words, the analyzed result with PLS reflects smaller deviations (noise) by sample because of nearly perfect fit. PLS is very promising, but we have to read the analyzing result carefully to decide the result that reflects the entire tendency or the particular sample.

One of the analyzing results is as follows. *Curative* garden should have 0 big standing stones, more than 6 middle-size stones aligned not linearly, 6 to 10 small stones and stepping stones; type is modern Japanese; one pine tree and tall stone lantern should be arranged in the garden.

5.7 Smoothing with Local Regression

There are several tabular data that are crucial for Kansei/affective engineering analyses. A design element table contains sample design variables and their variations. Kansei evaluation values are also made into a table. In Kansei/affective engineering, a design variable is called a design element. A design element table has several items. An item corresponds to a variable, such as color, shape of detail, illustration, layout of controls, and so on. One item has variations. For example, item body color has variations like black, silver, light blue, and dark blue. Variations are often called categories. The design element table has items and their variations on columns, and samples are allocated as rows (see Section 5.5). When a sample is painted dark blue on its body, a cell of [corresponding row, dark blue column] has value 1, and other cells of color variations have 0. In many Kansei/affective engineering cases, the design element table is larger.

The design element table has been used for analysis of the relationship between Kansei evaluation and design variations. QT1, which has been used widely in Kansei analysis, is a variant of a multiple regression that uses dummy variables to represent a qualitative (nominal scale) variable. If there are n types of variations of a nominal variable, $n - 1$ dummy variables are introduced. Although relationships between a Kansei word and design elements can be analyzed with QT1, the inside correlational structure of design elements is still unclear. Since relationships between columns and rows are discrete, tabular data are semilinear or nonlinear. Thus, new statistical analysis techniques for tabular data have been continuously developing in statistical science. In this section, we present an attempt to visualize evaluation values on a two-factor table.

5.7.1 Visualization of Lower-Dimensional Tabular Data of Design Elements and Kansei Evaluation

5.7.1.1 Parametric and Nonparametric Models

A first-order linear regression minimizes the error E over the entire data. The *a* and *b* are the regression coefficients, X_i is a value of the predictor (independent or explanatory) variable, and Y_i is a value of the target (dependent or explained) variable.

$$(\text{minimize})E = \sum_{i=1}^{n}(aX_i + b - Y_i)^2$$

A traditional regression tries to fit one regression expression to the entire domain of the existing data. Thus, the coefficients *a* and *b* have one value. The intent is to express the entire tendency with a simpler model by reducing parameters. This is called a parametric regression.

In contrast, nonparametric regression does not have to reduce the number of parameters (Takezawa, 2005). A nonparametric regression extracts useful tendencies in local features.

5.7.1.2 Local Linear Regression Model

A local linear regression (LLR) is a nonparametric technique. In most cases, an LLR tries to fit one regression expression to each set of existing data. Thus, each Y value is given different regression coefficients *a* and *b*. This is called *local fitting*. At one value of the predictor variable X_j, values for a_j and b_j can be derived where w_{ij} are weighting parameters that determine how much local data are used in the regression. For example, a Gaussian function can be used since it has a single peak and smooth tails.

$$(\text{minimize})E(X_j) = \sum_{i=1}^{n\_local} w_{ij}(a_j X_i + b_j - Y_i)^2$$

Finally, the estimated value of $\hat{Y}_j$ corresponding to X_j is

$$\hat{Y}_j = \hat{a}_j X_j + \hat{b}_j,$$

where the derived values of a_j and b_j are written with a "hat." Figure 5.29 shows an example of a locally fitted linear regression. Each line estimates one $\hat{Y}_j$ from X_j and near Xs with w_{ij}. The result shows the smoothing of Y.

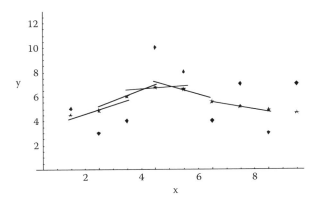

FIGURE 5.29
Locally fitted linear regression (dots are measured Y values).

We used LOESS (Cleveland et al., 1992), a local regression method, to smooth the data. Some methods use fixed interval sizes for local regressions, whereas LOESS uses k neighbor Xs of X_j. This idea achieves auto-tuning of smoothness: In the area that has many samples, smoothing is smaller; and in a sparse sample area, smoothing is larger. For the weighting of neighborhood X_j, LOESS uses a tricubic function given by

$$\left(1 - \left(\frac{|X_i - X_j|}{\text{distance to furthest neighbor}}\right)^3\right)^3$$

The span is defined as the control parameter for the size of the neighborhood area and is given as *k/number of data*. A larger value of the span gives more smoothed results and ignores local effects, whereas a smaller value of the span yields less smoothing but accounts for local effects.

LOESS and other local regression methods are now used for visualizing geographical data. A common example is mapping fishery investigation data. A region that was expecting a good catch was densely investigated. In other words, the sampling rate is relatively more often than other areas. Other regions that were not expecting a good catch were investigated sparsely. LOESS is particularly suitable for such an uneven sampling of the area.

5.7.2 Application to Hair Design Evaluation

LOESS smoothing has been applied to hair design Kansei evaluation data.

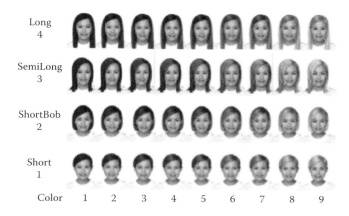

FIGURE 5.30
Hair design evaluation samples.

5.7.2.1 Kansei Evaluation Experiment

In the experiment, we used four lengths of hair: short, short bob, semilong, and long. Hair color has nine grades, from pitch black to gold blonde. Thus, the number of control variable combinations is 36 (4 × 9). Evaluation samples are made with hair design simulation software (Cosmopolitan Virtual Makeover 2, Broderbund, 1999). We chose a hairstyle that has fewer effects of cut and perm and is consistent shape and length as various preset hairstyles. Hair colors were chosen to make a continuous, natural change from commercially available hair coloring using the color simulation function of Virtual Makeover2. Seventy-six pairs of Kansei words were used for the questionnaire. Evaluation is rated on a 5-point SD scale. Stimuli were presented with a color-controlled display (iMacDV, Apple). Sixteen female college students, age 19 and 20, participated as subjects with no payment.

5.7.2.2 Analysis Method

Table 5.11 shows the result for the Kansei word *attractive*. Each cell has the number of subjects who rated 4 or 5 on the stimulus. Order of length and color is assigned the same as Figure 5.1. This table is the starting point of the analyses.

The upper graph of Figure 5.31 shows the two-dimensional plot of Table 5.11. Density of red (or dark in B/W) shows the number of subjects (high-density red cell has a larger number).

From Table 5.11 and Figure 5.31, we can see these facts. Color 4 (marrons glacés color) has a larger number. Many of the subjects agree that this hair coloring is attractive. On hair length 4, colors 2, 4, 5, and 6 have larger numbers.

TABLE 5.11

Frequency Table of *Attractive*

Length										
4	7	9	5	9	9	10	5	5	3	
3	4	6	4	9	6	5	6	4	5	
2	5	3	4	10	8	4	7	3	7	
1	3	4	7	9	7	5	4	5	7	
	1	2	3	4	5	6	7	8	9	Color

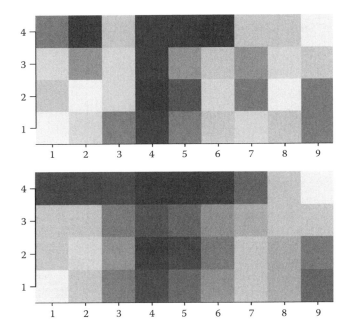

FIGURE 5.31
2-D plot, number of subjects who rated *attractive*. Nonsmoothed (upper), smoothed with local regression (lower) (denser is more frequent).

There are several difficult things. Middle-numbered cells such as 6 to 8 are scattered on tables. These cells make interpretation difficult. This table is unsuitable for analysis of variance (ANOVA), since there seems to be local interaction effects.

Smoothing methods have been used for data in an interval scale. Such examples are shown in the next section. Simonoff (1998) showed LOESS could be used for ordinal scale in one-dimensional data and in two-dimensional data (form of a table). We referred and modified an S program of Simonoff (1998). It uses LOESS implementation by Professor B. D. Ripley (Applied Statistics, Univ. of Oxford, UK). Computing platform is R-1.6.2 ported to Mac OS X by Professor Jan de Leeuw (UCLA).

The lower graph of Figure 5.30 shows the LOESS smoothing of *attractive*. Span was set to 0.25. There is a broad peak around color 3, 4, and length 2. In length 4, from black to brown, there is a loose, T-shaped, high-valued zone. We conclude that (1) longer hair is more attractive in a wide color range, and (2) from short to semilong, brown color is attractive. By smoothing small fluctuations, interpreting becomes uncomplicated.

Some other Kansei words have more linear relationships between color and length.

For the Kansei word *airily*, color is dominant. Brighter is more *airily*. Shorter hair has a slightly positive effect. For the Kansei word *chic*, color is also dominant. Darker color is more chic (Figure 5.32).

Visualizing the structure and interactions between design elements is also explained in the next section.

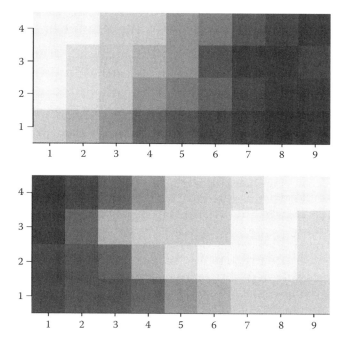

FIGURE 5.32
Smoothed *airily* (upper) and *chic* (lower).

5.8 Correspondence Analysis and Quantification Theory Type III

In the previous section, we described a method for visualizing complicated interactions between design elements on the two-dimensional table. Usually, Kansei/affective engineering studies have to analyze a much larger number of design elements. If a large design element table can be visualized, it is a great help for understanding the entire structure and details of design elements. In this section, we attempt to visualize larger design element tables. In this approach, all samples are mapped on a two-dimensional plane according to a statistical analysis of the design elements. Quantification theory Type III (QT3) and correspondence analysis are used for the analysis. Next, a three-dimensional contour map is created for the specific Kansei word. Kansei evaluation value for each sample is added as a height value augments the map. Then a smooth contour that interpolates between the Kansei values of the samples is computed by a local regression method. The proposed methodology creates a three-dimensional contour map that helps researchers to recognize both the linear and nonlinear relationship between a Kansei evaluation and the design variables. It consists of two stages: two-dimensional mapping of the design samples, and raising the height of the design samples in the map according to their Kansei evaluation values, which transforms the map into three dimensions. The heights are organized into contours that are connected and smoothed by a local regression method.

5.8.1 QT3

QT3, proposed by Hayashi (1954), is a method used to analyze the relationships within multivariate categorical data. Hayashi used the term *item* for a variable and *category* for the variations of a variable. For example, if a categorical variable is colors of shoes, there may be eight categories, such as red, pink, yellow, green, light blue, deep blue, black, and white (Tables 5.12 and 5.13). QT3 is a method of analyzing the relationship between categories of items where there are no other criteria. It could be considered a special case of a PCA.

The fundamental idea of QT3 is to give similar values y to samples having similar response patterns on categories, and to give similar values x to categories having similar response patterns on samples (Enkawa, 1988). Samples with similar response patterns on categories are collected together, and categories with similar response patterns on samples are also collected together. In other words, the rows and columns of the data table will be sorted so that similar responses are placed in orthogonal cells. Hayashi called this sorting procedure "seeking internal consistency."

If we regard the rows and columns as coordinates in a two-dimensional scatter plot, the sorting procedure becomes an estimation of x (weights of

TABLE 5.12

Example Data for QT3

Sample	Sneaker	Slip-On	Red	Pink	Light Blue	Deep Blue	Black	White
1	1	0	0	0	1	0	0	0
2	1	0	0	1	0	0	0	0
3	1	0	0	0	0	1	0	0
4	0	1	0	0	0	0	0	1
5	0	1	0	0	0	1	0	0
6	0	1	1	0	0	0	0	0
7	0	1	0	0	1	0	0	0
8	1	0	0	0	0	0	0	1
9	1	0	0	0	0	0	1	0
10	1	0	0	0	0	0	0	1
11	0	1	0	0	1	0	0	0

TABLE 5.13

Aligned Categories by QT3

Sample	Pink	Black	Sneaker	White	Deep Blue	Light Blue	Slip-On	Red
2	1	0	1	0	0	0	0	0
9	0	1	1	0	0	0	0	0
8	0	0	1	1	0	0	0	0
10	0	0	1	1	0	0	0	0
3	0	0	1	0	1	0	0	0
1	0	0	1	0	0	1	0	0
4	0	0	0	1	0	0	1	0
5	0	0	0	0	1	0	1	0
11	0	0	0	0	0	1	1	0
7	0	0	0	0	0	1	1	0
6	0	0	0	0	0	0	1	1

categories) and y (weights of samples) under the condition of maximizing the correlation between x and y. This is equivalent to the procedure for obtaining a weighted set of linear equations that maximize the variance. The procedure is the same as the computation of principal components in a PCA.

5.8.2 Correspondence Analysis

Correspondence analyses (CAs) have been studied since the early 1970s (Benzecri, 1992) as a method to visualize the relationship between rows and columns of a contingency table. The concept and procedure of a CA are based on that of Pearson's chi-square test, which is the most popular method to test the relationship between rows and columns in a table. In the chi-square test, the expected value of each cell is calculated from the sum of the row, the sum of the column, and the grand total. Thus, the expected value of the cell with subscript "*ij*" is (sum of the row *i*) × (sum of the column *j*)/ (*grand total*). The difference between the expected and the observed value is (*observed – expected*)^2/*expected*. The total of the difference over all the cells is the chi-square value. If this is larger than a certain value with respect to the number of degrees of freedom, the null hypothesis is rejected and we conclude that the rows are significantly associated with the columns.

The difference table is *i*-dimensional in rows and *j*-dimensional in columns. The rows and columns are projected onto a reduced two-dimensional space while performing the correspondence analysis. The rows and columns are also projected onto a reduced dimensional space. After normalizing and superimposing the two reduced dimensional spaces, we can visualize the relationships between items listed in the rows and columns of the difference table.

5.8.2.1 Correspondence Analysis Procedure

1. CA Step 1: Summation in the rows and the columns, and calculating the grand total. The column sums, the row sums, the grand total (53 in this example), and the "mass," which is the proportion of each of them in the grand total, are shown in Table 5.14.

2. CA Step 2: Calculating three tables below. Correspondence matrix **P** has the proportion of the original value in the grand total. The p_{ij}, a value in *i*th row and *j*th column, is given by $p_{ij} = original\_cell\_value/ grand\_total$. Each cell of the Row Profile Table has the value of p_{ij}/r_i. r_i is the *i*th row mass. By defining $\mathbf{D}_r$, the matrix that has row masses as diagonal elements, the Row Profile Table can be written in as $\mathbf{D}_r^{-1} \mathbf{P}$. Likewise, in the Column Profile Table each cell has the value of p_{ij}/c_j. c_j is the *j*th column mass. Matrix $\mathbf{D}_c$ has column masses as diagonal elements.

TABLE 5.14

Example Data for the Correspondence Analysis (Purchase of Shoes)

Age	Department Store	Given as Gift	Shoes Room Supermarket	Wagon Sale or Discount	Row Sum	Row Mass (*r*)
3 years old	6	3	4	1	14	0.264
4 years old	2	0	11	6	19	0.358
5 years old	2	0	17	1	20	0.377
Column Sum	10	3	32	8	53	
Column Mass (*c*)	0.189	0.057	0.604	0.151		

3. CA Step 3: Making the Standardized Residuals Table **A**. As its name shows, each cell has a standardized difference (a_{ij}) between the observed value and the expected value by

$$a_{ij} = (p_{ij} - r_i c_j) / \sqrt{r_i c_j}$$

4. CA Step 4: Singular value decomposition (SVD) of the table **A**. SVD is written in the next simple expression.

$$\mathbf{A} = \mathbf{U}\boldsymbol{\Gamma}\mathbf{V}^\mathrm{T}$$

The matrix $\boldsymbol{\Gamma}$ has singular values as its orthogonal elements. Squared singular values $\gamma_1^2, \_\gamma_2^2, _____\gamma_k^2$ are the eigenvalues of both $\mathbf{A}^\mathrm{T}\mathbf{A}$ and $\mathbf{A}\mathbf{A}^\mathrm{T}$. The column vector of **V** (the row vector of $\mathbf{V}^\mathrm{T}$) is the eigenvector of $\mathbf{A}^\mathrm{T}\mathbf{A}$. This vector is equivalent to that obtained by the common PCA on the matrix **A**. The column vector of **U** is the eigenvector of matrix $\mathbf{A}\mathbf{A}^\mathrm{T}$. This vector is equivalent to that obtained by the PCA on transposed matrix **A**, so-called row-mode PCA or Q-type analysis. Rows are treated as variables and columns are samples in this type of PCA.

5. CA Step 5: Mapping variables onto the reduced dimensional space. At first, the rows are projected by using its eigenvectors. The principal coordinates of rows (**F** in the form of matrix) are computed from u (the component of matrix **U**), singular value $\gamma\_$ and row mass r.

$$\mathbf{F} = \mathbf{D}_r^{-1/2}\,\mathbf{U}\boldsymbol{\Gamma}$$

In scalar notation: $f_{ik} = u_{ik}\,\gamma_k / \sqrt{r_i}$, where u_{ik} is the ith element of column vector $\mathbf{u}_k$, which is the kth eigenvector of the PCA of transposed **A**. The i corresponds to the ith row in the original table. Thus, f_{21} is the second element on the second variable (the second row) of the first eigenvector. A row of the original table is expressed with its eigenvector, which has adjusted with corresponding square-rooted eigenvalue and its row mass. Second, the columns are also projected by eigenvectors. The principal coordinates of columns (those from matrix $\boldsymbol{\Gamma}$) are computed with the following equation:

$$\mathbf{G} = \mathbf{D}_c^{-1/2}\,\mathbf{V}\boldsymbol{\Gamma}.$$

A column is expressed with an eigenvector, which has adjusted square-rooted eigenvalues and its column mass. Now, rows and columns each have a map.

6. CA Step 6: Standardization of coordinates for making the plane square to map both the rows and the columns on the same plane. Standardized coordinates of rows are provided by the following equation. The principal coordinates of rows are divided by their respective singular values:

$$X = F\Gamma^{-1}$$

The standardized coordinates of columns are also provided by the following equation. Principal coordinates of columns are also divided by their respective singular values.

$$Y = G\Gamma^{-1}$$

5.8.2.2 Correspondence Analysis Example

We summarize the procedure of a correspondence analysis with the data of the parents' purchase patterns of shoes for children. Details of the evaluation will be mentioned later. We found 53 valid responses for the ages of the children and purchase points.

Figure 5.33 shows the results of the CA of the data contained in Table 5.14, where *3 years old* is close to *given as present, 4 years old* is close to *wagon sale or discount*, and *5 years old* is relatively close to *shoe room supermarket*.

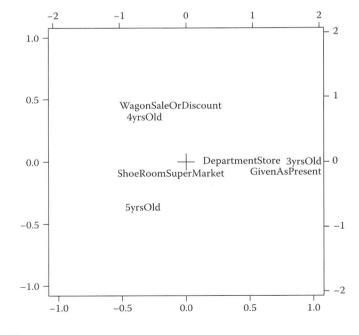

FIGURE 5.33
Correspondence analysis example of shoe purchase data.

```
> library(MASS)
> cadata <- read.table("CAexample.txt", header=T, row.names=1)
>biplot(corresp(cadata,nf=min(nrow(cadata),ncol(cadata))-
1),xlim=c(-1.0,1.0),ylim=c(-1.0,1.0))
```

5.8.3 Solving the QT3 Model with the CA

Some researchers have shown that the QT3 method is a special case of the CA. For example, Yanai (1994) described the theoretical and computational similarities between these two methods and demonstrated that the QT3 model can be solved within the framework of the CA. CA software packages have been developed with recent algorithm advancements, and these can be used to solve the QT3 model. The results of the present analysis are shown in Figure 5.34 for shoe types and Table 5.12 for colors.

```
> itemCategoryTable <- read.table("QT3example.txt",header=T)
> biplot(corresp(itemCategoryTable,nf=min(nrow(itemCategoryTab
le), ncol(itemCategoryTable))-1))
```

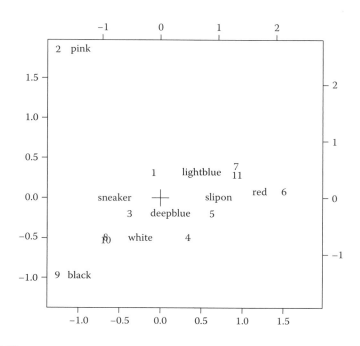

FIGURE 5.34
QT3 model analyzed with CA, shoe types and color.

5.8.4 Application to Children's Shoes Study

5.8.4.1 Evaluation Experiment

Twenty-three mothers evaluated the Kansei words evoked by 29 samples of children's shoes that were collected from major shoe manufacturers in Japan. The participants evaluated each pair of shoes on 5-point SD scales for 31 Kansei words.

5.8.4.1.1 Map of the Design Samples

The design samples were mapped onto a plane by solving the QT3 model with a CA. There were 52 categories in the design variables. Figure 5.35 shows a dense map of the samples and variables in which the samples are allocated in an arch-like shape, giving a slight horseshoe effect. Some researchers who use correspondence analyses argue that the second dimension is not important because it is merely a sort of quadratic function of the first dimension (Gifi, 1990). Some effort has been directed toward removing the horseshoe effect from the CA (Hill and Gauch, 1980). The results shown in Figure 5.35 do not have a problem due to the horseshoe effect as a result of our interpretation of the Kansei evaluation. The following sections describe the implication of Kansei words for both dimensions.

In this case, design elements have a 29-dimension space. Dimensions of the designs were reduced from 29 to 2 with QT3 computation. Design samples were mapped on a plane projected from the highest number of dimensions. In this case, the dimensions of the designs were reduced from 29 to 2. Then, the correlations between the coordinates and the original variables of each sample were computed to evaluate the major patterns in the design space. Because the procedure of computing the correlations was similar to the procedure of obtaining loadings in a PCA or factor analysis, we called the correlation values *loadings*.

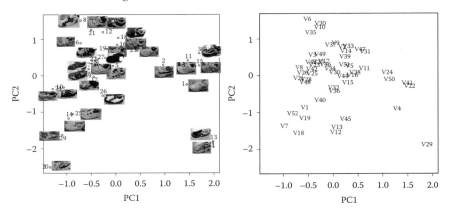

FIGURE 5.35
Map of samples (upper) and map of design elements (lower) by QT3.

Along the first axis, the design variables that had correlations greater than +0.5 were sneakers (V5), thick-heeled (V41), wavy-shaped soles (V29), zippers (V4), heart-shaped printings (V22), and shiny ornaments (V24). The design variables that had values less than −0.5 were slip-ons (V1), thin-heeled (V42), soft soles (V48), and no shiny ornaments (V25). The most characteristic design element along the first axis was thick-heeled (V41) on the right side and thin-heeled (V42) on the left side.

The design variables that had correlations greater than +0.5 along the second axis were Velcro (V2, V3) and curved toes (V39). Typical shoes with these variables were designed for infants aged 1 or 2 years. The variables that had correlations less than −0.5 were lightweight (V45), slip-ons (V7), no parts for adjustment (V1), noncurved toes (V40), and light blue colored (V12). Typical shoes with these variables were inexpensive and designed for 4- to 5-year-old boys and girls.

In addition to the allocation of the general design tendency, we found several specific shoe groups on the map. Sneakers for girls with thick heels (V41), hard-soled (V51), and shiny ornaments (V22, 24) were on the right side and almost above the second axis (samples 3, 11, 15, and 1). Girls' wave-patterned thick-soled (V29) sneakers with zippers (samples 13 and 4) were on the lower right. At the top center, infant (young toddler) shoes were mapped together (samples 8, 21, 12, 18, 16, and 23) in the area around zero on axis 1 and greater than 0.6 on axis 2. These consisted of sneakers with round-shaped toes (V39), large heel guards (V37), and Velcro for attachment (V2). Sporty sneakers with lines on the sides and few or no ornaments were located around the center (samples 6, 19, 27, 22, 28, 5, 7, 9, 17, and 26). These shoes shared many design variables. Red sneakers (V8) for girls (samples 24 and 10) were on the left-most side. Boys' slip-ons (V7) with thin soles (samples 25, 14, 29, and 20) were on the lower left. Since they were slip-ons, they had no adjustments (V1) and the inner pads were not changeable (V52).

5.8.4.1.2 Details in the Mapping of Samples

In addition to the allocation of the general design tendency, we found several specific shoes groups on the map. Sneakers for girls with thick heels (V41), hard soled (V51), and shiny ornaments (V22, 24) were on the right side and almost above the second axis (samples 3, 11, 15, and 1). Girls' wave-patterned thick-soled (V29) sneakers with zippers (samples 13 and 4) were on the lower right. At the top center, infant (young toddler) shoes were mapped together (samples 8, 21, 12, 18, 16, and 23) in the area around zero on axis 1 and greater than 0.6 on axis 2. These consisted of sneakers with round-shaped toes (V39), large heel guards (V37), and Velcro for attachment (V2). Sporty sneakers with lines on the sides and few or no ornaments were located around the center (samples 6, 19, 27, 22, 28, 5, 7, 9, 17, and 26). These shoes shared many design variables. Red sneakers (V8) for girls (samples 24 and 10) were on the left-most side. Boys, slip-ons (V7) with thin soles (samples 25, 14, 29, and 20) were on the lower left. Since they were slip-ons, they had no adjustments (V1) and the inner pads were not changeable (V52).

5.8.5 Contour Maps According to Specific Kansei Words with LOESS

We obtained different patterns of contours generated from the evaluation values, with LOESS, which was explained in Section 5.7. Some Kansei words yielded linear-sloped contours, whereas others produced local peaks. In all the results, we set the span equal to 0.5.

5.8.5.1 Linear Kansei Words

An evaluation of *fashionable* produced a linear change on the design element map. The design variables on the right consisted of thick-soled sneakers with shiny ornaments, which were evaluated as being more fashionable than thin-soled slip-ons. Conversely, *easy to walk* and *flexible* were more on the left.

Premium has high evaluations on the positive side on the second component. Velcro is more premium and slip-on is less premium (Figure 5.36).

5.8.5.2 Nonlinear Kansei Words

An evaluation of *showy* produced a basin along axis 2 and between 0 and –1 on axis 1. *Simple* resulted in an inverted contour of *showy*. The basin of *showy* was the hill of *simple*. These two Kansei words shared the most design variables. There was a difference in color: *Showy* shoes were two-colored, and *simple* shoes were one single color (Figure 5.37).

Evaluations for *individual* and *modern* both showed a peak around the center and a ridge from the center to the right of the map. *Clean* had an inverted

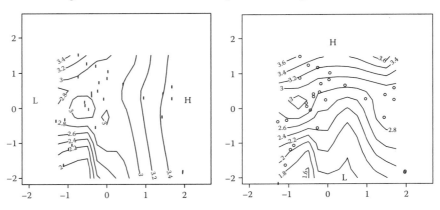

FIGURE 5.36
Contour map of linear Kansei according to design element map, fashionable (left) and premium (right).

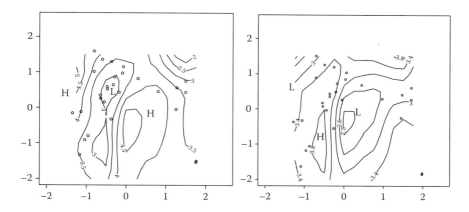

FIGURE 5.37
Contour map of nonlinear Kansei according to design element map, *showy* (left) and *simple* (right).

peak that was a basin around the center. Shoes rated as *individual* and *clean* shared shoelaces and heel guards. The difference in their design was that *individual* shoes had shoelaces and were two-colored, and *clean* shoes had both shoelaces and zippers and were a single color (Figure 5.38).

The Kansei words that commonly produced a hill on the upper left were *good material, lightweight, cushioned, stable, anti-slip,* and *sporty.* The common features among these Kansei words were a Velcro belt, no printings, curved toes, and mid-weight.

We also found other correspondences. Thick-soled sneakers with shoelaces were on the upper right region, which was modestly high in the *good shaped.* No shoes were rated above 3.5. The region between the left center and lower

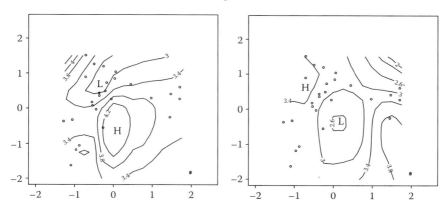

FIGURE 5.38
Contour map of nonlinear Kansei according to design element map, *individual* (left) and *clean* (right).

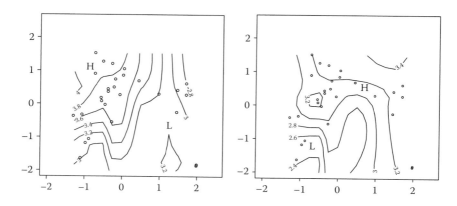

FIGURE 5.39
Contour map of nonlinear Kansei according to design element map, *good material* (left) and *well designed* (right).

left was high for *quickly dried*. Sneakers without any ornaments and slip-ons with thin soles were included in this region. *Cute* and *girly* were located at the lower right and associated with thick-soled sneakers for girls. Thin-soled slip-ons located at the lower left were associated with *easy to wash, easy to get into,* and *easy for the parent to put on*. *Stain resistant* had a peak at the center. *Well designed* had an inverted L-shaped ridge from the top to the right, with no peak above 3.5. The rest, which were not well-designed shoes, were thin-soled slip-ons for boys.

We proposed two methods for visualizing design element structures and relationship between Kansei evaluations. The first method is an attempt to visualize the two-factors evaluation experiment shown in Section 5.7. This approach utilizes local regression smoothing to show relationships between design elements and Kansei evaluation.

In this section, the second method aims to obtain three-dimensional contour maps to graphically show the relationship between Kansei words and design variables. These maps can help designers grasp whether a specific Kansei word is associated with a comprehensive design structure or with detailed variables. The former type of association produced a set of linear contours while the latter produced complex nonlinear-shaped contours. Because the contour maps can be plotted for every Kansei word, designers can examine the effects of both comprehensive and detailed designs.

References

Benzécri, J.P. (1992). *Correspondence analysis handbook*, Marcel Dekker, New York.
Brereton, R. (2003). *Chemometrics*, Wiley, Chichester, England.

Cleveland, W.S. , Grosse, E., Shyu, M. (1992). Local regression models, in Chambers, J., Hastie, T. (Eds.), *Statistical Models in S*, 309–376, Chapman & Hall, Boca Raton, FL.

Enkawa, T. (1988). *Multivariate analysis*, Asakura-shoten, Tokyo.

Gifi, A. (1990). *Nonlinear multivariate analysis*, Wiley, Hoboken, NJ.

Hayashi, C. (1952). On the prediction of phenomena from qualitative data and the quantification of qualitative data from the mathematical-statistical point of view, *Annals of the Institute of Statistical Mathematics*, **2**(3), 69–98.

Hayashi, C. (1954). Multidimensional quantification with applications to the analysis of social phenomena, *Annals of the Institute of Statistical Mathematics*, **2**(5), 121–143.

Hill, M.O., Gauch, Jr., H.G. (1980). Detrended correspondence analysis: An improved ordination technique, *Vegetatio*, **42**, 47–58.

Ishihara, S., Ishihara, K., Nagamachi, M., Nishino, T. (2001a). Mathematical modeling of nonlinearity on form ratio Kansei evaluation data, *Proceedings of the Fourth International Quality Management and Organizational Development Conference*, Linköping, Sweden, 536–542.

Ishihara, S., Ishihara, K., Nagamachi, M., Nishino, T., Komatsu, K. (2001b). An analysis of nonlinearity characteristics on Kansei data, *Systems, Social and Internationalization Design Aspects of Human-Computer Interaction*, **2**, 320–324, CRC Press.

Ishihara, S., Ishihara, K., Nagamachi, M., Nishino, T. (2003). Smoothing of ordinal categorical data and its application to analysis of 2-dimensional data of hair design Kansei evaluation data, *Proceedings of International Ergonomics Association Congress* (CD-ROM).

Ishihara, S., Komatsu, K., Nagamachi, M., Ishihara, K., Nishino, T. (2003). An analysis on nonlinearity of Kansei evaluation data, *Journal of Human Interface Society*, **5**(2), 267–274.

Ishihara, S., Nagamachi, M., Ishihara, K. (2007). Analyzing Kansei and design elements relations with PLS, *Proceedings of the First European Conference on Affective Design and Kansei Engineering* (CD-ROM), Lund University Press.

Kaufmann, L., Rousseeuw, P.J. (1990). *Finding groups in data*, Wiley Interscience, New York.

Mitsuchi, S. (1997). From the beginning of multivariate analyses, *Nippon Hyoronsha*. Tokyo.

Miyashita, Y., Sasaki, S. (1995). *Chemometrics-chemical pattern analysis and multivariate analysis*, Kyouritsu Publishing, Tokyo.

Romesburg, H.C. (1989). *Cluster analysis for researchers*. Robert E. Krieger Publishing, Malaber, FL.

Simonoff, J.S. (1998). *Smoothing methods in statistics*, Springer-Verlag, Heidelberg.

Sneath, P.H.A., Sokal, R.R. (1973). Numerical taxonomy, W.H. Freeman and Co., San Francisco.

Takezawa, K. (2005). *Introduction to Nonparametric Regression*, John Wiley & Sons, Hoboken, NJ.

Venables, W.N., Ripley, B.D. (1999). *Modern applied statistics with S-PLUS*, 3rd ed., Springer, Heidelberg.

Yanai, H. (1994). *Multivariate data analysis methods—Theories and applications*, Asakura-shoten, Tokyo.

6

Soft Computing System for Kansei/Affective Engineering

Yukihiro Matsubara

CONTENTS

6.1 Artificial Intelligence Technology and Kansei/Affective Engineering System

Kansei/affective engineering (KAE) is an effective technique for translating the human Kansei (consumers' feeling and desire for the domain product) into the product design element (Nagamachi, 1989). Recently, this technique has been implemented in many fields of product development. Actually, the concept of KAE is referred to in many phases of the product designing process. Furthermore, the Kansei/affective engineering system (KAES) functions as the interface between a product designer and a product consumer.

There are two kinds of techniques for KAE: (1) the forward inference type of KAE (from Kansei to the design element) and (2) the backward inference type of KAE (from a candidate design to diagnosed Kansei) (Nagamachi, 1995). Implementing these inference algorithms on the computer system, KAES acts as the expert system based on artificial intelligence technology (Matsubara et al., 1994a). In general, the forward inference type of KAES (called forward KAES) is utilized to support the consumer's decision in selecting the desired product. The backward inference type of KAES (called backward KAES) is utilized to support the designer's creative work diagnosing the Kansei regarding the designer's rough sketch. We proposed an intelligent image processing mechanism that utilizes the KAES, and a system that can recognize the designer's idea as the combination of design elements (Matsubara et al., 1994b). Furthermore, the combined computerized system of the forward and backward KAES would be a more powerful supporting tool for both consumer and designer.

In this section, we describe the hybrid KAES as the general framework of KAES, which is the combined forward KAES and backward KAES. First, we describe the support for consumer decision making and show the concept of hybrid KAES. Second, we explain the Kansei inference model, which is based on the linear regression model. Third, we show the detailed description of the hybrid KAES structure and design recognition function. Finally, we describe the prototype system for the hybrid KAES as the domain for the front door of a house, and conclude by evaluating the system.

6.1.1 Kansei/Affective Engineering and Decision Supporting

1. *Consumer's decision supporting.* Suppose a consumer wants to buy a house. He/she has a Kansei or feeling concerning his/her desire for the product. For example, he/she wants to construct a *luxurious* house at some price. They sit in front of the KAES computer and input their desired Kansei words into the KAES. It understands their desire through the inference engine using the databases and produces the final decisions from the computer, which matches their desire for the products. The KAES helps the consumer choose a product (Nagamachi, 1993, 1994a,b).

2. *Designer supporting system.* KAES is also used in designing a new product. When a designer is creating a new product, he/she starts with his/her product-designing image or concept. Then he/she consults with the KAES by inputting the designer's image words. The KAES outputs the calculated results through KAE on the display. If the displayed candidates are different from the designer's image, they can change the shape design and color by the KAES change procedure (Nagamachi, 1995).

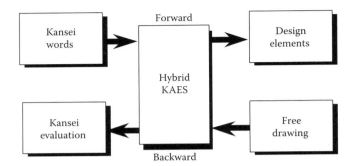

FIGURE 6.1
Diagram of hybrid KAES.

3. *Framework of hybrid KAES.* Hybrid KAES consists of forward KAES and backward KAES. Figure 6.1 shows the diagram of the hybrid KAES. Forward KAES is the KAES in which a designer obtains the desired design through an input of the Kansei word. In backward KAES, the designer is able to draw a rough sketch in the computer and the computer system recognizes the pattern of the design input by the designer. Then the system estimates the Kansei or image of the input design through the backward inference engine and shows the estimated level of desired Kansei about the design. The following functions are required for hybrid KAES.

   a. Forward Kansei inference mechanism
   b. Backward Kansei inference mechanism
   c. KAES database building supporting mechanism

6.1.2 Kansei Inference Model

We proposed some Kansei models that are based on both linear and nonlinear models to demonstrate the relationship between the human Kansei and product design element (Ishihara et al., 1995a,b; Manabe et al., 1994; Tsuchiya et al., 1994). In this chapter, we assumed the linear regression model as the Kansei inference model (see Figures 6.2 and 6.3) and formalized the following equation. This model is the typical linear regression model and can be analyzed and identified using Hayashi's quantification theory Type I (QT1) (Hayashi, 1976). This method is one of the categorical multiple regression analysis methods. The criterion variable is quantitative and the explanatory variable is qualitative (which means categorical parameter). First, we define the dummy variable $\delta i(jk)$ as follows.

$$\delta i(jk) = 1, \text{ when a sample } i \text{ corresponds to item } j \text{ and category } k \quad (6.1)$$

$$\delta i(jk) = 0, \text{ otherwise} \quad (6.1a)$$

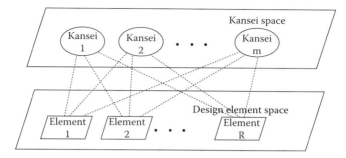

FIGURE 6.2
Relationship between Kansei and design.

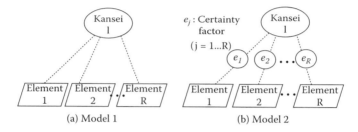

(a) Model 1   (b) Model 2

FIGURE 6.3
Kansei inference model.

where

$i = 1, \ldots, n$ (n: number of samples), $j = 1, \ldots, R$ (R: number of item),
$k = 1, \ldots, Cj$ (Cj: number of category for item j),

$$RR = \sum_{j=1}^{R} C_j \qquad (6.1b)$$

Using these dummy variables, suppose the linear regression model for the categorical explanatory variable in equation (6.2).

$$Yi = \sum_{j=1}^{R} \sum_{k=1}^{Cj} a_{jk} \delta_i(jk) \qquad (6.2)$$

In order to satisfy and minimize the following equation, apply the method of least squares for the criterion variable $\{y_i\}$ to identify the category score a_{jk}.

$$Q = \sum_{i=1}^{n} (yi - Yi)^2 \rightarrow \min \qquad (6.3)$$

Next, doing the partial differentiation for {Q} by each a_{jk}, we can get the RR case of linear equations. Solving these simultaneous equations, we can get the category score a^*_{jk} shown in equation (6.4).

$$Y_i^{(l)} = \bar{y}^{(l)} + \sum_{j=1}^{R}\sum_{k=1}^{Cj} a_{jk}^{*}{}^{(l)}\delta_i(jk) \qquad (6.4)$$

Now, we assume the $Y_i^{(l)}$ as the evaluated value of the specific Kansei l ($l = 1$, ..., m; number of Kansei words), we can say that $a^*_{jk}(l)$ indicates the relationship between the specific Kansei l and the design element corresponding to item j, and category k.

6.1.3 Hybrid KAE

6.1.3.1 Overview

Figure 6.4 shows the system structure of hybrid KAES. This system consists of four main modules (design processing module, inference module, Kansei word processing module, and system controller), and five kinds of database (design database, graphic database, knowledge base, image database, and Kansei word database). When the user (either consumer or designer) inputs the Kansei word by natural language, the system tries to identify the Kansei meanings through the Kansei word processing unit referring to the Kansei word database. Then the system infers the candidate design through the forward inference engine referring to the knowledge base and image database. Finally, the system outputs and displays the candidate design using CG (computer graphics) through the picture drawing module referring to the design database and graphic database (see Figure 6.5a).

When the user inputs the combination sets of design elements, the system performs the design element recognition module and identifies the design elements as the item and category. If the user inputs the free-drawing picture, the system uses the image processing and recognition techniques and can get the identified results (Matsubara, et al., 1994a,b). Then the system outputs the diagnosis results, which are the Kansei or images concerned with the inputted design through the backward inference engine and explanation processing unit (see Figure 6.5b). Next, we show an explanation for each forward and backward inference engine mechanism in detail.

6.1.3.2 Forward Inference Engine

Suppose the user inputs and requests the Kansei l^*. We already obtained the equation (6.4) concerned with the Kansei l^* and identified each a^*_{jk}, which is stored in image database. Then the system can infer the adequate category k^* concerned with Kansei l^* performing the following steps for each item.

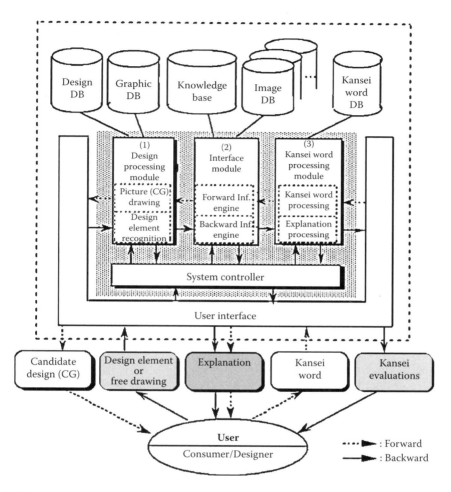

FIGURE 6.4
The system structure of hybrid KAES.

Step 1. START, $j = 1$.

Step 2. $a^*_{jk^*} = \max_k[a^*_{jk}]$.

Step 3. item j: select the category k^*.

Step 4. $j = R \rightarrow$ Step 5; otherwise $\rightarrow j = j + 1$, Step 2.

Step 5. Referencing the knowledge base (expert knowledge) if needed, modify the design element adequately (if there are conflict conditions between each item and category).

Step 6. Decide the candidate design, END.

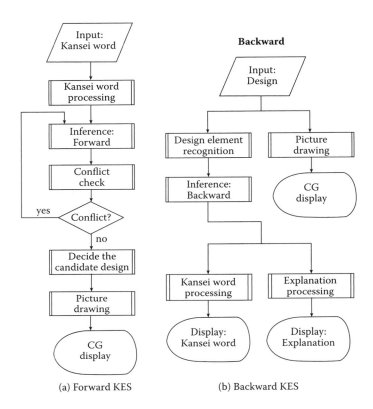

(a) Forward KES   (b) Backward KES

FIGURE 6.5
The system flow.

6.1.3.3 Backward Inference Engine

Set the following equation based on the equation (6.4).

$$S^{(l)} = \sum_{j=1}^{R} e_j^{(l)} \sum_{k=1}^{cj} a *{}_{jk}^{(l)} \delta_i(jk) + c^{(l)} \tag{6.5}$$

where: $e_j(l)$: certainty factor, $c(l)$: constant term.

If the user inputs the specific candidate design i^*, the system can identify each item and category corresponding to input design. As the results, identify $\delta^* i(jk)$, then the system can get $S(l)$.

$$e_j^{(l)} = 1 \qquad (j = 1, ..., R) \tag{6.6}$$

When putting the $e_j(l)$ in equation (6.6), the system actualizes the model 1 shown in Figure 6.3a. If we assume an adequate certainty factor $e_j(l)$ (e.g., the specific value which is introduced for referring the partial correlation coefficient (PCC) calculated by quantification theory Type 1), we can actualize the model 2 shown in Figure 6.3b. Furthermore, we can get the $S(l)^*$ as follows in equation (6.7):

$$S^{(l)}* = \frac{S^{(l)} - \min_{j,k}\left[S^{(l)}\right]}{\max_{j,k}\left[S^{(l)}\right] - \min_{j,k}\left[S^{(l)}\right]} \tag{6.7}$$

We define the $S(l)^*$ as the fitness score for Kansei l corresponding to input candidate design i^*. Iterating the above procedure for all l, we can get the diagnosis result for backward Kansei inference. In this section, the system has four types of inference procedures as the combination of establishment method for $e_j(l)$ and $a^*_{jk}(l)$ as follows.

1. $e_j(l) \leftarrow 1$, $a^*_{jk}(l) \leftarrow$ normal
2. $e_j(l) \leftarrow 1$, $a^*_{jk}(l) \leftarrow$ standardized
3. $e_j(l) \leftarrow$ referring to the PCC, $a^*_{jk}(l) \leftarrow$ normal
4. $e_j(l) \leftarrow$ referring to the PCC, $a^*_{jk}(l) \leftarrow$ standardized

6.1.4 Design Element Recognition Subsystem

This section focuses on the "front door of a house" as the design element to recognize. The housing environment is one of the most important fields for expressing personal taste and feelings. Each product unit is so expensive that it is difficult to change them frequently. Therefore, the customer is asked to make careful decisions regarding the product unit. The customer desires a decision support tool to realize and translate his own demand. Hybrid KAES is a very useful tool for supporting the consumer's decision as well as the designer's use. The front door is a good example of a design element. Furthermore, it is easy to recognize because the shape is a simple part of a square and circle.

6.1.4.1 The Definition of Design Element Recognition and Identification

We give the definition of recognition used in KAES as "the attempt to express by the variables (parameter set of item and category) for the input design picture." The item is an attribute that constitutes a design. For example, in the case of a front door, there are door colors, door frames, transoms, door structures, crosspieces, and so forth (see Table 6.1). A clear definition exists

TABLE 6.1

Classification of Design Elements for Front Door of House

Item			Category
Frame	Door Frame	(1*)	Alone, One Side Wing, Both Side Wing, Double Door
	Transom	(2*)	Nothing, Door Transom, Frame Transom
Color	Door Color		White, Gray, Black, Pastel, Brown
Material	Material		Wood, Other
Lattice	Lattice	(11*)	Nothing, Width Rows, Matrix, Diagonal Cross, Other
Crosspiece	Door Structure	(10*)	Normal Top Crosspiece, Stone-Bridge Type, Semicircular Type, With One Point, Flush Door 1, Flush Door 2
	Width Crosspiece	(4*)	Nothing, Balanced One, Lower One, Upper One, Balanced Two, Lower Two, Upper Two, Over
	Lengthwise Crosspiece	(3*)	Nothing, One, Two
	† Type Crossings of Crosspiece	(5*)	Nothing, One, Two, Over
	††Type Crossings of Crosspiece	(6*)	Nothing, One, Two, Over
	Inside Short Crosspiece 1	(8*)	Nothing, Existence
	Inside Short Crosspiece 2	(9*)	Nothing, Existence
	T-Crosspiece	(7*)	Nothing, T Type, A Type, Both B Type, Type, Other

for each item. Therefore, the recognition algorithm must be constructed referring to this item definition. There are some clear relations among the items, so we use this relation structure in the recognition algorithm. One relation is that the transom exists in an upper part of the door. When searching for the transom on the input picture, the system remembers this rule and carries out the recognition for the upper area of the picture.

6.1.4.2 The Recognition Algorithm

The system gets the portrait information by an image scanner and extracts an outline of the object, then tries to divide the portrait referring to the relation structure rule (Design database). Next, the system constructs the hierarchical structure of the design element using the recognition algorithm.

6.1.4.3 Recognition of the Outside Door Frame Structure

The flow chart to recognize the outside frame (1*) is shown in Figure 6.6. Using the relation structure rule, the system tries to judge whether and how

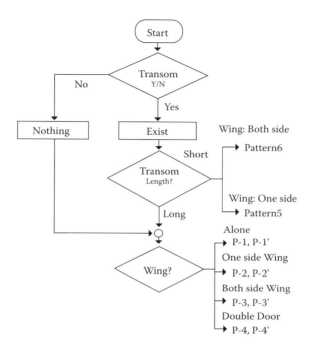

FIGURE 6.6
The flow of recognition algorithm.

the transom exists (2*) on the alone-type door and gets a result for which type of specific pattern to choose among the 10 patterns shown in Figure 6.7.

6.1.4.4 Recognition of Alone-Type Door

The detailed recognition for the alone-type door is carried out by the next step.

Step 1. Distinction of Length Crosspiece and Width Crosspiece
Item: Lengthwise Crosspiece (3*), Width Crosspiece (4*), Single-Type Crossing Crosspiece (5*), Double-Type Crossing Crosspiece (6*), Door Structure (10*)

Step 2. Distinction of Outside Frame and Inner Crosspiece
Item: Inside short Crosspiece 1 (8*), Inside short Crosspiece 2 (9*)

Step 3. Distinction of T-character Crosspiece
Item: T-character Crosspiece (7*)

Step 4. Distinction of Lattice
Item: Lattice (11*)

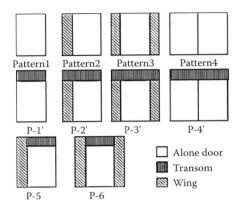

FIGURE 6.7
The 10 patterns of door structure.

In Step 1, the system counts the number of crosspieces. In Step 2 and Step 3, this algorithm identifies the shape of the crosspiece chosen in Step 1. In Step 4, the system infers the internal structure of the outside frame using the mask processing technique and statistics method.

6.1.5 Application: Front Door of House

6.1.5.1 Preparation

We prepare some data to construct the system that performs the following basic steps (Nagamachi, 1989):

1. Selection of the adjective words (Kansei words). Collect many adjectives words that have a relation to the object.
2. Experiment: Assess slides or pictures on SD scales. Make pairs of these adjectives in a good—bad fashion for the SD (Osgood's semantic differential) scales, and then make an image questionnaire for Kansei experiments. Next, the subjects assess many samples (slides or pictures) related to the object on these SD scales.
3. Identification of factor analytical structure. Calculate the assessed data at Step 2 by factor analysis, and obtain the semantic factorial structure of adjectives on the related design. Elicit adjectives that have a close relation to the object from the collected adjectives.
4. Assessment of the design components. Subdivide the object design on the slides or pictures into the detail design components, and then classify each component into a category according to its quality.
5. Multivariate analysis of estimated data. Using the assessed data at Step 2 and qualitative data at Step 4, analyze by Hayashi's quantification theory Type 1, which is a type of multivariate regression

analysis dealing with qualitative data. The results of this analysis show relevancy between an adjective (criterion variable) and each design component (explanatory variable).

6. Development of KAES. We can construct the Kansei database for KAES according to the result of Step 5.

6.1.5.2 Kansei Experiment and Statistical Analysis

6.1.5.2.1 Kansei Experiment

The image questionnaire is a basic trial that verifies the relationships between human Kansei and the design elements. A Kansei database can be built on its results. In an image questionnaire, specialists subjectively evaluate some sample designs by SD scales of Kansei words, consisting of pairs of adjectives and their antonyms. The notation of design elements and attribute notation of item/categories are adopted. Sample designs are selected for reference. In this case, some experimental conditions are the number of Kansei words, item/categories, sample designs, and human experts for the front door of the house are defined as follows:

1. Forty pairs of adjectives are selected as the Kansei words. Useful pairs were selected from 800 adjectives from front door catalogs and magazines.

2. Thirteen kinds of attributes (item/categories) are selected for the design component, such as door frame, door color, lattice, and so on, of a front door. Categories are a few values for each item. In total, we have 54 categories (see Table 6.1).

3. Eighty-two types of front doors are selected as sample designs.

4. Seventy-seven people joined the experiment as the human experts.

6.1.5.2.2 Statistical Analysis

The data obtained by the image questionnaire were analyzed statistically with QT1. The data from the image questionnaire can be considered trustworthy because the multiple correlation coefficients (MCCs) of the statistical analysis are distributed between 0.80 and 0.95. In relation to this, see the Kansei word pair *beautiful—ugly* (Table 6.2) as an example of the results. The MCC is 0.8874, which is the medium value for the range of the MCCs. Viewing the partial correlation coefficients (PCCs), those of *door color, transom*, and *width crosspieces* are highest of all. Therefore, it can be suggested that these items affect the Kansei word. Further, seeing category scores of these items, for *door color*, white doors or gray doors are beautiful, and pastel doors or brown doors are ugly; for *width crosspieces*, lower two or upper two doors are beautiful; and for *transom*, a transom door is ugly.

TABLE 6.2

Result of Quantification Theory Type I Analysis of *Beautiful*

ITEM	PCC	CATEGORY	CATEGORY SCORE: BEAUTIFUL -UGLY
Door Frame	0.366	Alone	−0.0269
		One Side Wing	0.0754
		Both Side Wing	−0.0101
		Double Door	−0.2249
Door Color	0.744 (1)	White	−0.3013
		Gray	−0.1921
		Black	0.0018
		Pastel	0.2962
		Brown	0.2395
Transom	0.520 (2)	Nothing	−0.0933
		Door Transom	0.2165
		Frame Transom	−0.0755
Door Structure	0.452 (5)	Normal Top Cros.	0.015
		Stone-Brid. Type	0.044
		Semicircular Type	0.402
		With One Point	−0.1424
		Flush Door 1	−0.265
		Flush Door 2	0.0202
Material	0.010	Wood	−0.0086
		Other	0.0006
Width Crosspiece	0.517 (3)	Nothing	0.0596
		Balanced One	−0.1084
		Lower One	−0.0973
		Upper One	−0.1527
		Balanced Two	0.2006
		Lower Two	−0.2396
		Upper Two	−0.4119
Lengthwise Crosspiece	0.499 (4)	Over	0.0843
		Nothing	0.1278
		One	0.0516
		Two	−0.2715
Lattice	0.389	Nothing	0.0513
		Width Rows	0.0328
		Matrix	−0.004
		Diagonal Cross	−0.0246
		Other	−0.2157
+ Type Crossings of Crosspiece	0.278	Nothing	−0.0165
		One	−0.0685
		Two	0.2864
		Over	0.1433
++ Type Crossings of Crosspiece	0.387	Nothing	−0.0485
		One	0.3729
		Two	0.2386
		Over	0.0256
Inside Short Crosspiece 1	0.153	Nothing	0.0073
		Existence	−0.2929
Inside Short Crosspiece 2	0.203	Nothing	−0.0139
		Existence	0.2701
T - Crosspiece	0.318	Nothing	0.0001
		T Type	0.0556
		T Type	0.1062
		Both Type	−0.2854
		T Type	0.0548
		Other	0.1894

(Axis scale: −0.6, −0.3, 0.3, 0.6)

6.1.5.3 Building the Kansei Database

The Kansei database is built to depend on the data obtained by the image questionnaire and the statistical analysis, reflecting the knowledge of specialists. There are two ways to describe the relationship between the Kansei words and the design elements. The first one is composed of two databases: (1) an Image database describing item weights and category scores for each Kansei word, and (2) a Basic Design database describing constraint conditions for some designs. The advantage is that a Kansei database can be easily tuned and renewed because item weights and element scores are made to meet PCCs and category scores of the statistical analysis. In this case, however, the expert system must have a consistent reasoning process. The second way is heuristic, in which candidate designs have been decided beforehand and describe each Kansei word. In this way the expert system necessitates little reasoning, but the candidate design for each Kansei word will be definite. In this case we built a Kansei database according to the first way and will consider improvements for the database and reasoning method in the future.

6.1.5.4 Forward Inference

Figure 6.8 shows a sample run of the expert system for front door designing. This system is implemented on an Apple Macintosh using the programming language C. In running the system, at first a window for adjective selection is displayed as in Figure 6.8a, the user chooses one or more adjectives, and the system runs the reasoning. Then the system draws a picture as an output, as Figure 6.8b shows. After this, the user can change the detail designs of the output designs.

6.1.5.5 Backward Inference

The output example of backward Kansei evaluation results is shown in Figure 6.9. The recognition result (two kinds of pictures and the list of recognized items) is indicated in the upper part of the figure. The reasoning result for the Kansei is shown in the lower right part, and a menu button to make detail changes is indicated in the lower left part.

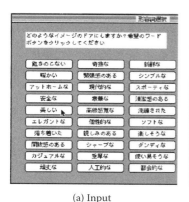
(a) Input

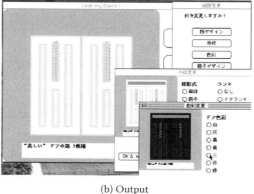
(b) Output

FIGURE 6.8
An example: Forward.

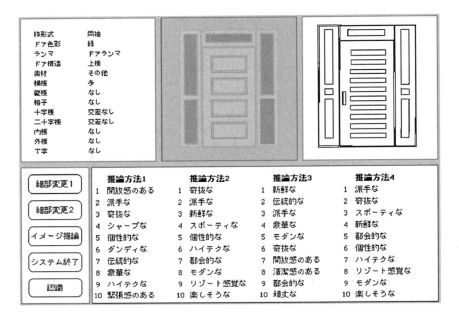

FIGURE 6.9
An example: Backward.

6.1.5.6 Design Recognition

We collect the 82 pattern design samples that exist in the marketing catalog and estimate the recognition rate for the design element recognition subsystem to try recognizing those samples (Matsubara et al., 1994b,c). The system divides the picture into the door frame level as Hierarchy-1, the shape of the crosspiece level as Hierarchy-2, and the texture of lattice level as Hierarchy-3 (see Figure 6.10). Each hierarchy's recognition rates are shown in Figure 6.11. In this result, high hierarchy has good performance in recognizing the element. This is the reason why the lower hierarchy received the influence for the higher one's recognition error. Therefore, it is able to improve the rate to increase the sufficient recognition rules.

6.1.6 Summary

In this section, we described the concept of KAES and explained hybrid KAES. First, we described the function of the support tool for customer decision making and showed the concept of hybrid KAES. Then we explained the Kansei inference model, which is based on the linear regression model.

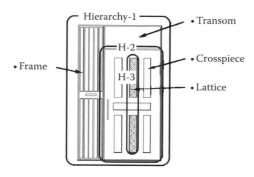

FIGURE 6.10
The hierarchical structure for the door.

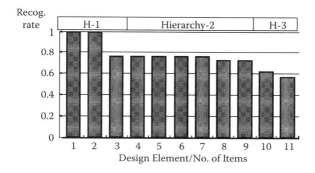

FIGURE 6.11
The recognition rate for each item.

Finally, we showed the detailed description of hybrid KAES structure and design recognition function, and explained the prototype system for hybrid KAES as the domain for the front door of a house.

This section has treated the front door of a house as an object domain. This is an object for easy recognition. However, the shape is more complex in a design like the car interior. A usual design element has a curve in its shapes, and there are continuous characteristics between each design element. Therefore, it is difficult to apply this algorithm simply to other domains. We are now proving the new algorithm to recognize such complex design elements based on the neural network model (Neocognitoron) (Kashiwagi et al., 1994; Jindo et al., 1994). Using this approach, we can apply hybrid KAES to another more general domain.

6.2 Neural Network Model

6.2.1 Types of Neural Network

Neural network models have been well explored and developed from the late 1980s to the 1990s. From the viewpoint of a learning method of the neural network, most of them are roughly divided into two: networks for supervised learning and networks for unsupervised learning.

Supervised learning is to learn rules with the training data. The network inducts the relations between input and the correspondingly desired output. In practical usage, supervised learning network dominates in the most of the applications. Major network structures for supervised learning are multi-layered perceptron (MLP) with an algorithm for error back propagation, recurrent networks (most of them are MLP with "context" units), and radial basis function networks.

Unsupervised learning is a categorization process without any feedback on whether the result is right or wrong from out of the network. This kind of network only receives input samples; right outputs corresponding to the input are not provided.

There are three major classifications of neural networks that perform unsupervised learning. The first group is interconnected networks. Typical ones are brain-state-in-a-box (BSB) networks of Hopfield and J.A. Anderson. BSB is similar to an analog version of the Hopfield model. Interconnected networks have fully interconnected neurons, and they act as autoassociative memory machines. In other words, they are content-addressable memory systems; they can recall the entire pattern from a part of it.

The second group is networks with Hebbian learning rule. The most practical network of this kind is PCAnet, which performs principal component analysis (PCA). The application of PCAnet in KAE will be discussed later.

The third group is competitive learning networks. Kohonen's network for learning vector quantization and self-organizing map (SOM) is well known. Another practical network is the adaptive resonance theory (ART) network proposed by Grossberg and Carpenter. We have modified and utilized ART-type networks for many cases in KAE.

6.2.2 PCAnet and Its Kansei Engineering Applications

We show mechanisms and properties of PCAnet and examples of KAE applications (Ishihara, 1995, 1997). We assume a neuron y_1 that receives input signals from N neurons $x_1, x_2, \ldots, x_N$. The c_{1i} is a synapse weight from x_i to y_1. Output signal of y_1 is defined as:

$$y_1 = \sum_{i=1}^{N} c_{1i} x_i \qquad (6.8)$$

The classical Hebbian learning rule updates the synapse weights as shown in equation (6.9).

$$c_{ji}(t+1) - c_{ji}(t) + \gamma y_j(t) x_i(t) \qquad (\gamma: \text{constant}) \qquad (6.9)$$

The following is an explanation for why the Hebbian rule maximizes the variance of output vector: When updating synapse weights, frequently input patterns add similar values to c_{1j}, and come to have a large influence on y_1. Suppose that the inputs have a positive correlation between x_1 and x_2. In this case, most inputs have the same sign for both x_1 and x_2 ([+,+] or [−,−]). As learning progresses, y_1 comes to have a large value when both x_1 and x_2 have large values with the same sign ([+,+] or [−,−]). Thus, the variance of y_1 is maximized in the direction to maximize the correlation between x_1 and x_2. This process is similar to the procedure of eigenvector extraction from a correlation matrix in PCA. Each component of c can infinitely grow in the original Hebbian rule. Oja proposed adding feedbacks to the network to converge the sum of squares of each c to 1. He also proved that $c_{11} \ldots c_{1N}$ become the first eigenvectors when converged (Oja, 1982).

6.2.2.1 Generalized Hebbian Algorithm

We describe Oja's algorithm and the extension for extracting the second and succeeding eigenvectors. Equation 6.10 shows Oja's algorithm that finds only the first eigenvector that has the largest eigenvalue. Thus, equation 6.10 is used in the case when there is a single output (y_1) neuron.

$$c_i(t+1) = c_i(t) + \gamma y(t) \left[x_i(t) - y(t) c_i(t) \right] \qquad (6.10)$$

Sanger (1989) extended the Hebbian learning rule constrained by Oja's algorithm and called it the generalized Hebbian algorithm (GHA). Here, we define a single-layer network that is constructed by a single layer of processing neurons $\mathbf{y} = \mathbf{Cx}$, where $\mathbf{x}$ is an n-dimensional input vector, $\mathbf{C}$ is an $m \times n$ weight matrix, $\mathbf{y}$ is an m-dimensional output vector with $m < n$, and γ is a rate of changing the weights. In this chapter, we regard a network that implements GHA as a PCAnet. The architecture of PCAnet is shown in Figure 6.12. A correlation matrix of the input is defined as $\mathbf{Q} = E[\mathbf{xx}^T]$. GHA is described as:

$$c_{ji}(t+1) = c_{ji}(t) + \gamma(t) \left(y_j(t) x_i(t) - y_j(t) \sum_{k \leq j} c_{ki}(t) \, y_k(t) \right) \qquad (6.11)$$

Equation (6.11) is a modification rule for the synapse weight between the ith element of the input vector and the jth neuron. The synapse weights and

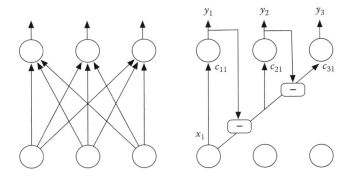

FIGURE 6.12
Structure and learning rule of PCAnet.

previous outputs from (1, …, (j–1)th) neurons negatively affect the modification of the jth neuron's weight. GHA is build by combining Oja's algorithm
and the Gram-Schmidt orthogonalization algorithm. It extracts m eigenvectors successively.

In this algorithm, if we maintain the diagonal elements of CC^T equal to 1,
then a Hebbian learning rule makes the rows of C converge to the m eigenvectors of Q. Thus, $CC^T = I$ and $Q = C^TAC$, where A is the diagonal matrix of
eigenvalues of Q in descending order. The weight adaptation process guar-
antees $CC^T = I$. GHA provides a practical procedure for finding m eigenvectors without calculating Q.

Figure 6.13 (left) shows a trajectory of c in the orthogonalization process. In
this example, the center of distribution of the set of input data (shown as dots)
is [0,0]. The PCAnet in this case has two neurons that receive two-dimen-
sional input vectors. After 200 inputs, $c_{11} = 0.82$, $c_{12} = 0.63$, $c_{21} = -0.22$, $c_{22} =
0.62$ (shown as solid arrows). Two y neurons have mutually orthogonalized
synapse weights that correspond to the first and the second eigenvectors.

Figure 6.13 (right) shows the orthogonalization process on actual Kansei
evaluation data. Seventy-five photos of suits were evaluated by five female
college students. Evaluation was done with 73 Kansei words. PCAnet was
provided each subject's evaluation data, successively for every sample, from
sample 1 to sample 75. Synapse weights' corresponding eigenvectors were
plotted. To make a comparison, the standard method of PCA by QR method
was also shown. Figure 6.13 (right) shows x axis values of the 1st eigenvector
and 2nd eigenvector. With 40 sample presentations, eigenvectors are stable.

This result implies that PCAnet is suitable for online learning. Recently, we
have obtained data continuously from databases, such as customer purchase
data. The traditional QR method requires a large correlation matrix, and it
requires entire recomputation for each addendum of the data. Thus, PCAnet
is very promising as an online data analysis tool, and it can deal with both
small data and huge data for data mining.

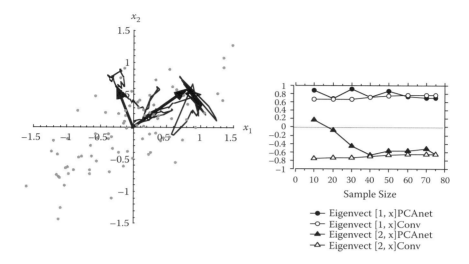

FIGURE 6.13

Orthogonalization by PCAnet: 2-dimensional data (left) and 73-dimensional Kansei evaluation data.

6.2.2.2 Application of PCAnet to KAE

We show an example of KAE evaluation data analysis with PCAnet (Ishihara et al., 1995). We selected 40 female watches and presented them to 18 female college students. Forty-four adjective word pairs were used for evaluation. Five-point SD scale evaluations were encoded into 0.25 step values between 0 and 1.

PCAnet computes eigenvectors and principal components (PCs) of the data. (Initial value of γ is 0.04, 10 iterations of all data were presented.) Three major PCs are extracted. PC loadings of each adjective are calculated from variance of output of the PCAnet. Figure 6.14 shows Kansei words that have large (+/−) PC loadings on three PCs. PC1 is *soft–hard*. At the positive side of PC1 are *soft, tender,* and *feminine*. On the opposite side of PC1 are *gloomy, hard,* and *tough*. PC2 represents a dimension of appeal. The negative side (closer side) has *plain, dark,* and *calm,* and the positive side (farther side) has *showy, gorgeous,* and *elegant*. PC3 represents activity. The positive side (upper) has *sporty, immature,* and *casual,* and the negative side (lower) has *adult, intellectual,* and *calm*.

Output vectors of PCAnet are PC scores of each product sample. Figure 6.15 shows PC scores on the main three PCs. Numbers on the three-dimensional graph correspond to the sample number. Pictures shown are typical examples that have large (+/−) PC score on each PC. For PC1, a sample of large positive PC score has a white oval face and a light-brown leather belt. This watch corresponds to *tender, soft,* and *feminine*. A sample with a large negative value on PC1 has a white square face and belt made of metal. This watch corresponds

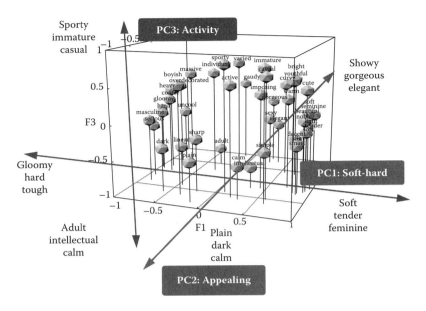

FIGURE 6.14
Principal component loadings by PCAnet (watch data).

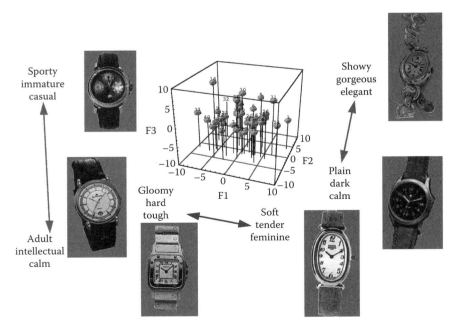

FIGURE 6.15
Principal component scores by PCAnet (watch data).

to *gloomy, hard,* and *tough.* On PC2, the watch with a black simple face and brown leather belt corresponds to *plain* and *dark.* A sample that corresponds with *showy, gorgeous,* and *elegant* is a gold bracelet-type. On PC3, the most sporty watch is a leather belt chronograph, and most *adult, intellectual,* and *calm* watches were classically designed with gold case and a white panel.

6.3 ArboART: Self-Organizing Neural Network–Based Hierarchical Clustering

6.3.1 Competitive Learning and Self-Organizational Clustering

Competitive learning performed by a neural network is a kind of unsupervised learning. The network discovers the characteristics of input data, such as typical patterns, general attributes, correlations, or categories. Thus, units and connections between units of the network are required to be somewhat self-organized. That is accomplished under the same situation as that in cluster analysis, since it classifies input data without correct answer patterns.

The objective of the competitive learning network is classification of input data. Similar inputs have to be categorized into the same group (cluster); that is, the same unit has to fire when similar data are inputted. As mentioned above, which cluster to be categorized into is found by the network itself according to the similarities among input data.

The simplest form of a network for competitive learning has an output layer and an input layer. An input unit x_i in the input layer connected is connected with all output units (one of the output units is denoted o_j) by the excitatory connection weight $w_{ji} \geq 0$. We show an example structure of such a network, where a black dot receives each dimension of input vector ζ, a white circle represents a output unit, an arrow is an excitatory link, and the symbol "$-|$" represents an inhibitory link.

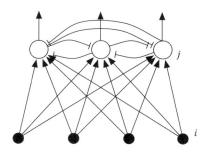

FIGURE 6.16
Simple competitive learning neural network.

A single output unit fires at a time. This unit is called a winner unit. It receives the largest amount of input in general. The amount of input T_j that the jth output unit o_j receives according to input vector ξ is:

$$T_j = \sum_i w_{ji}\xi_i = \mathbf{w}_j \cdot \xi \qquad (6.12)$$

In the case that the weight vectors are normalized (e.g., $|\mathbf{w}_j| = 1$ for all j's), the winner unit J is decided by:

$$\mathbf{w}_J \cdot \xi \geq \mathbf{w}_j \cdot \xi \quad \text{(for all other } j) \qquad (6.13)$$

where the weight vector $\mathbf{w}_J$ becomes the closest to the input vector ζ.

Generally, the lateral inhibition process of neurons is omitted in the description and the operation of an artificial neural network; the decision process of a winner is described as the choice of an output unit that receives the maximum input within the network ($\mathbf{w}_j \cdot \xi$) or ($\mathbf{w}_j \cdot \mathbf{x}$).

A neural network that learns by competition is able to learn so that it categorizes its input data, where the initial values of connection weights w are appropriately randomized. Here, we geometrically explain the behavior of a general neural network for competitive learning, following Hertz, Krogh, and Palmer (1991). Refer to Figure 6.17.

Suppose a case of three-dimensional data and an input vector $\mathbf{x} = (x_1, x_2, x_3)$. Each input vector $\mathbf{x}$ is represented by a dot placed on a unit sphere, whose center is at $(0, 0, 0)$. A weight vector of jth output unit from each input $\mathbf{w}_j = (w_{j1}, w_{j2}, w_{j3})$ is represented as a symbol "×" on the sphere, where the weights are normalized as $|\mathbf{w}_j| = 1$.

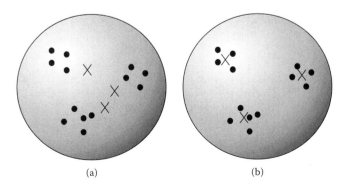

(a)   (b)

FIGURE 6.17
Finding clusters by competitive learning.

When an input is given to a network, a winner is decided according to equation (6.13). The winner unit (×) is one that has the closest weight vector to the input data (•). Then, the weight of the winner unit J, $\mathbf{w}_J$ is updated according to the learning rule. The learning rule varies with the type of the network.

In this way, the closest unit to the input data is chosen, and then its weight vector is pulled toward the center of the distribution of input data. In Figure 6.17, (a) an initial state and (b) a typical state at the completion of learning are shown. When learning is completed, output units find clusters of input data. Each output unit changes its weights toward the center of the cluster it represents. After all, we can find clusters of input vectors using competitive learning by a neural network.

A theory of competitive learning is applied to different form of networks. A self-organizing network (Kohonen, 1989) and variations of ART networks (Carpenter and Grossberg, 1987, 1991) are good examples.

6.3.2 ART1.5-SSS

In this chapter, we first describe the basic mechanism of ART1.5-SSS (small sample size), which is a building block of arboART.

ART-type neural networks are based on the adaptive resonance theory proposed by Grossberg and Carpenter. Their attempt, when developing the theory, was to create a stable self-organizing system in a changing environment. ART-type neural networks have features like unsupervised learning and a self-organizing pattern classifier. They added a reset mechanism, which compares an input vector with the cluster prototype according to the distance criteria. ART is built by adding the reset mechanism to a simple competitive neural network that can make stable clusters and maintains plasticity to new inputs.

There are several versions of ART-type neural networks. ART1 handles 0/1 digital vectors. ART2 deals with continuous, analog vectors (Carpenter and Grossberg, 1991). ART1.5 (Levine and Penz, 1990) is a simplified version of ART2, where the normalized process of input vector is removed. ART3 has a closer mechanism to actual neuron cells' electrochemical signal passing. Since Kansei evaluation data have continuous value and the range of each variable (evaluation value) is narrow enough, we have chosen ART1.5 as a prototype mechanism for our classifier. We have modified its learning rule and named it ART1.5-SSS.

6.3.2.1 Structure of ART1.5-SSS

F2 units are interconnected by inhibitory links. Suppose that x_i is the activity of ith F1 unit, and T_j is activity of jth F2 unit. z_{ji} is the bottom-up connection weight from the ith F1 unit to the jth F2 unit. Reversely, z_{ij} is the top-down connection weight from *j*th F2 unit to the *i*th F1 unit. The initial value of the bottom-up connection z_{ji} is set as small random number around 0. The top-down connection z_{ij} is set as 0.

The F1 unit outputs x_i just like the input signal ξ is.

$$x_i = \xi_i \qquad (6.14)$$

$$T_j = \sum_i z_{ji} x_i \qquad (6.15)$$

In the competition process, an F2 unit that receives the maximum input signal multiplied by bottom-up connection weights is activated. T. means all of T units.

$$T_j = T_j : \text{if } T_j = \max \left(T_* \right)$$

$$T_j = 0 : \text{otherwise} \qquad (6.16)$$

A reset mechanism sends a reset signal to the unit in the F2 layer when its prototype and the input vector are not similar. x is the vector of F1 activities (x_i), and z_j is the vector of top-down weights (z_{ij}) from the *J*th F2 unit (the winner). When the following equation is satisfied, a match occurs. Otherwise, a reset signal is sent, and a search occurs. r is a constant of angle threshold called a vigilance parameter.

$$\frac{(\mathbf{x} \cdot \mathbf{z}_J)}{|\mathbf{x}||\mathbf{z}_J|} > r \qquad (6.17)$$

When a search occurs, the algorithm selects the unit that outputs the next maximum and tests it again. If all committed nodes (F2 units that are used for representing categories) failed the test, an uncommitted node is chosen for a new category. The cluster generation process is stabilized by this reset and search mechanism.

Connection weights between the chosen unit in F2 and all the units in F1 are modified so that they come slightly closer to the input vector. We can interpret the top-down connection weights from an F2 unit (that represents a category) to F1 units and bottom-up connection weight vectors from F1 units to the F2 unit (the category) as a prototype vector of input vectors that was assigned to the category. In ART1.5, a top-down weight z_{ij} and a bottom-up weight z_{ji} are updated by equations (6.18) and (6.19), where a is a learning rate, a constant between 0 and 1. We use a different learning rule from this in our ART1.5-SSS.

$$\frac{d}{dt}z_{Ji} = a\left\{-(1-a)z_{Ji} + x_i\right\}$$ (6.18)

$$\frac{d}{dt}z_{iJ} = a\left\{-(1-a)z_{iJ} + x_i\right\}$$ (6.19)

ART-type neural networks perform self-organizing clustering in the process mentioned above.

6.3.2.2 ART1.5-SSS: Improvement of ART1.5 for Small Sample Size

ART1.5-SSS is a revised version of ART1.5 that the authors have developed (Ishihara et al., 1993, 1994b, 1995a) to ensure accurate clustering under small sample size conditions. *Small sample size* means that the number of samples is not large in relation to the dimensions of the attributes (Hamamoto et al., 1994). It is a common practical problem in the case of supervised pattern classification (Raudys and Jain, 1991).

In the learning rule of the original ART1.5 proposed by Levine and Penz (1990), all input vectors assigned to a category equally affect their prototype. This means that the category prototype is changed throughout the clustering session. Consequent categories may include some old members that are rather different from recently assigned members.

Ill-defined clusters are often produced in situations as follows: When an input sample is categorized, the category's prototype is adjusted toward to

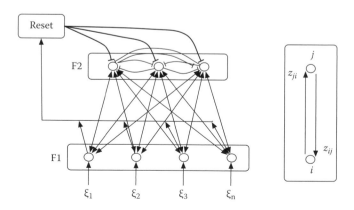

FIGURE 6.18
Structure of ART1.5-SSS.

the new input. A fixed learning rate causes continued shift of the prototype vector in pursuit of new input samples. Therefore, the boundary of a cluster also moves along the prototype shifts. As categorizing and learning proceed, a category boundary might overlap other categories. Consequently, category boundaries of these clusters may become difficult to interpret and early-categorized members may be different from new members and final proto-types; similar samples might be divided into different clusters, or a cluster would contain dissimilar samples.

With a constant rate of learning, meaningless clustering often occurs in that only a few members are assigned to each category. A large sample of natural data is normally distributed in general, but a small-sized sample is not always evenly distributed.

The original learning rule with a fixed learning rate works well when the input vectors (samples) are much more than the categories. In such cases, a fixed learning rate makes a prototype vector close to the center of the nor-mally distributed data.

However, in most cases of Kansei engineering, usually the number of products is not so many compared to number of the clusters. The original learning rule may not work appropriately in such cases.

Thus, we modified the learning rule of ART1.5 so that it can control the influence of the newly categorized input vector to the prototype.

This modification allows the prototype vector to contain traces of the pre-viously categorized vectors. This idea has come from a learning function in the algorithm of MacQueen's (1967) adaptive k-means clustering, which is used as a traditional (nonneural) pattern classification technique.

We use the following equations (6.20) and (6.21) in place of equations (6.18) and (6.19):

$$\frac{d}{dt} z_{Ji} = \frac{1}{q_J}\left(x_i - z_{Ji}\right) \tag{6.20}$$

$$\frac{d}{dt} z_{iJ} = \frac{1}{q_J}\left(x_i - z_{iJ}\right) \tag{6.21}$$

where, q_J represents the number of times the Jth F2 unit is chosen.

The new rule decreases the revising rate of the connection weights in fre-quently committed categories to avoid excessive changes of prototypes and to keep the characteristics of early classified samples in the cluster. We can expect reasonable clusters even in small-sized sample conditions with this improvement. We have shown the improvement in clustering small-sized samples in Ishihara et al. (1993, 1995a).

6.3.3 Hierarchical Clustering Using ART Networks: arboART

We have developed hierarchical clustering algorithm arboART (Ishihara et al., 1995b, 1999a, 2005). The basic idea of arboART is cascading clustering using multiple ART networks. A prototype formed in an ART network is used as an input to other ART networks that have looser distance criteria. For example, let us consider the case of hierarchical clustering using three different layers, ARTa, ARTb, and ARTc, as shown in Figure 6.19. The vigilance parameter r is the distance criterion used in ARTa, which is the strictest distance criterion of the three.

First, all samples are processed by ARTa, which may use rather a lot of category units, where each of them has a few (sometimes only one) samples. We note these n clusters made by ARTa as $cl(a_1, \ldots, a_n)$. After the one-shot learning, a top-down weight vector from each category unit becomes a prototype of its category. We use all top-down weight vectors as input vectors to ARTb.

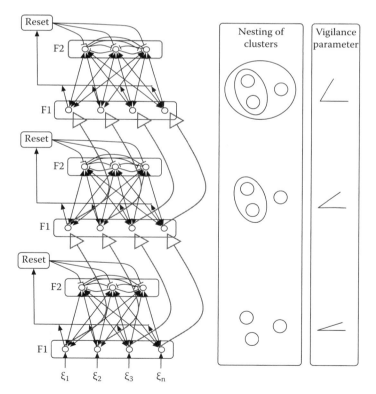

FIGURE 6.19
Structure of arboART.

ARTb makes superclusters of those made by ARTa. It has looser distance criteria than ARTa; thus, some clusters of cl(a_1, ..., a_n) are merged into a new cluster, and others remain. We represent m, fewer clusters made by ARTb as cl(b_1, ..., b_m).

ARTc in the third takes top-down vectors of cl(b_1, ..., b_m) as its input vectors. Its distance criterion is looser than that of ARTb. Therefore, ARTc merges some of the clusters cl(b_1, ..., b_m) and builds s, fewer clusters cl(c_1, ..., c_s).

Many small categories are combined into small numbers of larger (more generalized) categories at the end of iteration so that the prototype vectors built in early ART are cascaded to other ARTs with looser distance measure. All samples can be classified into one cluster through a long cascade.

Therefore, we can draw a dendrogram using the classification records of sample and categories. We call this idea arboART, where *arbo* means *tree* in Esperanto.

This algorithm for the hierarchical clustering is described as follows:

1. Initialization:

   Set all the elements of top-down vectors in all ARTs to 0.

   Set all the elements of bottom-up vectors in all ARTs to 1/(square root of the dimension of input vectors). If the dimension of input vectors is less than 10, however, set them to a square root of 10, to avoid too-large initial values for a few dimensions.

   Set the vigilance parameter r of ARTa to r_a, where $0 \leq r_a < 1$.

   Set the step st between an ART and the following one to a moderate value. The smaller st is, the more layers of ART that are required and the more detailed the hierarchy is derived.

2. Iteration of clustering in ARTs:

   In the case of first process, set $r = r_a$, otherwise, set $r = r$ (previous ART) $- st$.

   Drive the ART and send the obtained category prototype vectors to the next ART to build superclusters.

   Repeat this step until all the samples are merged into a single category.

3. Draw a dendrogram based on the record of categorization collected from each ART.

4. Find the features of a category from its prototype vectors.

The advantage of this ART-based hierarchical clustering is less cost of recalculation when new data are added. In case some evaluation samples are added to the experiment, iterative calculations are required in conventional methods for entirely creating a larger similarity matrix. On the other hand, ART-based computation does not handle the similarities in the form of a matrix; input samples are directly classified into a category. Additional inputs are also directly classified; that is all that is required.

6.3.3.1 Validation of arboART's Clustering Accuracy

In Ishihara et al. (1999), we validated clustering ability of arboART with standard test data set of StatLog project (Michie et al., 1994). We compared arboART clustering with traditional methods: WPGMA (weighted pair-group method using arithmetic averages), UPGMA (unweighted pair-group method using arithmetic averages), CLINK (complete linkage clustering method), weighted centroid, centroid, and SLINK (single linkage clustering method). The comparison was using *heart* data. Heart data have 13 attributes such as age, sex, symptom, various biochemical values, and absence/presence of heart disease in 270 subjects. We divided 270 heart samples into nine parts, after the StatLog method (Michie et al., 1994). Thus, there are nine trials, and each trial used different 30 samples. We made nine trials as one set, and we made another two sets. In another two sets, we randomly shuffled samples to avoid ordering effect of samples. Then, finally, we performed 27 (nine trials by three sets) trials.

In this study, we use the definition of clustering error that was used for comparative study of conventional clustering methods (Bayne et al., 1980).

cls is the previously obtained class label of each sample. Since the test data we used are two classes problem, each data labelled as class 1 or 2, thus *cls* is 1 or 2. v_{cls} is the number of the samples that belong to the class *cls*. When a sample belonging to class *cls* was classified into class *cls'*, it is written as (*cls'*|*cls*). Suppose we obtained NC clusters. C_i is one of the clusters. $m_i(cls'|cls)$ is the number of *cls*-labeled samples belonging to cluster C_i.

We can define that the cluster C_i belongs to class 1 when $m_i(cls'|1)/v_1 > m_i(cls'|2)/v_2$, and otherwise to class 2. The cluster is judged as belonging to one class, which is the class of the majority of members. Since the number of the member differs by the class in the heart data set, it should be balanced by dividing v_1 or v_2.

The error rate is defined as follows. When C_i was defined as belonging to class 1, $mi(1|1)$, the number of samples, which have class label 1 at C_i, is the number of correctly classified samples. Then, $mi(1|2)$, the number of samples which have class label 2 at C_i, is the number of classification error.

The total sum of all clusters' classification error at a clustering trial is NE.

$$NE = \sum_{i=1}^{NC} \{m_i(1\,|\,2) + m_i(2\,|\,1)\} \qquad (6.22)$$

Another criterion is inner-cluster error, which is used for measurement of clustering appropriateness (Dubes and Jain, 1976). A smaller cluster is generally preferred. A larger cluster tends to include not-so-similar samples. Inner-cluster error is calculated from the sum of differences between each member of a cluster (ξ_r) and center of cluster c_i. $n\_of\_member_i$ is a number of the C_i member. E^2 is a total sum of squared inner-cluster error of clusters.

$$c_i = \frac{1}{n\_of\_member_i} \sum_{r \in C_i} \xi_r \qquad (6.23)$$

$$E^2 = \sum_{i=1}^{NC} \sum_{r \in C_i} (\xi_r - c_i)^T (\xi_r - c_i) \qquad (6.24)$$

In hierarchical clustering, cluster number is reduced by combining clusters. Thus, we cut the dendrogram (tree graph of cluster combining process) at a stage with the appointed number of clusters and then we measure error at those clusters. The cutting point of the dendrogram is at 14-cluster solution, which is half of the number of samples. We used arboART with eight stages of ART1.5-SSS. Vigilance parameter of the first stage r_a is 0.95. The difference between next stage st is +0.05. In the case of absence of 14 clusters solution (in some set), st was modified within +−0.03, only at nearest cluster stage.

We show classification error NE and inner-cluster error E^2 in Table 6.3. Figures that have statistical significant differences are indicated with a star. For NE, arboART has the best result. SD is also the best. For E^2, arboART is the second best. SD is also the second.

For classification error NE, we tested the difference between clustering methods with one-factor analysis of variance (ANOVA). The difference was significant ($F(6,182) = 4.319$, $p = 0.004$). Multiple comparisons using Tukey's test shows arboART's NE is significantly smaller than weighted centroid and nonweighted centroid (MSe = 2.549, p < 0.01).

For inner-cluster error E2, we also tested the difference between clustering methods with one-factor ANOVA. The difference was significant

TABLE 6.3

Average Number and Standard Deviation of Classification Error and Inner Cluster Error Values of Various Clustering Methods

	NE (average)	NE (SD)	E^2 (Average)	E^2 (SD)	No Chaining
arboART	2.963	1.315	4.182	0.698	OK
WPGMA	3.222	1.601	4.870*	0.824	OK
UPGMA	3.333	1.569	4.778	0.812	OK
CLINK	3.630	1.573	5.116**	0.948	OK
W-Centroid	4.481**	1.827	7.845**	1.180	OK
Centroid	4.556**	1.847	7.848**	1.180	OK
SLINK	3.222	1.368	3.576	0.653	NG

* $p < .05$; ** $p < .01$

(F(6,182) = 92.722, p = 0.001). Multiple comparison using Tukey's test shows arboART's E^2 is significantly smaller than weighted centroid, nonweighted centroid, CLINK (MSe = 0.847, p < 0.01), and WPGMA (MSe = 0.847, p < 0.05). SLINK has smaller E^2 than arboART, but the difference was not significant.

We also examined nesting of clusters. In hierarchical clustering, generating intermediate clusters that combine subordinate clusters is commonly required. These intermediate clusters are useful for understanding hierarchical structure.

CLINK is often blamed for causing a *chain effect* when combining clusters. In such case, there are almost no intermediate clusters and each sample is merged sequentially to a few clusters. As a result, we could not see hierarchical relations between intermediate clusters (Sneath and Sokal, 1973).

In this validation, the SLINK method caused a chaining effect and often made one large cluster. The reason for smaller E^2 is that almost all samples except the large cluster are made into singleton clusters.

From these results, we believe that arboART produces excellent classification.

6.3.4 Application Example of arboART

6.3.4.1 *Kansei Evaluation Experiment*

In the research for Milbon's hair treatment development, which was mentioned in Chapter 2, we conducted an experiment for evaluating hair treatment packages. Forty-three samples of hair treatment containers were collected and were used for Kansei evaluation experiment. Fourteen subjects participated, female college students, ages 19 and 20. Thirty-nine adjective words (Kansei words) were used for evaluation using the 5-grade SD scale. Each subject picked up the sample hand and marked her impression. Evaluation took about 2 hours.

6.3.4.2 *Cluster Analysis by arboART*

We used arboART for package Kansei analysis. Input vectors for arboART were the averaged value of evaluations between subjects. Thus, there are 43 input vectors ξ, with 39 dimensions of Kansei evaluations. Evaluation values were averaged between subjects.

Each element of the input vector corresponded to an evaluation of each Kansei word, so the input vector had 39 elements, and there were 43 input vectors. We used arboART consisting of four layers of ART1.5-SSSs. As shown in Figure 6.20, various samples are hierarchically grouped by pattern of subjects' response to products. Enclosures in the figure correspond to nth ART1.5-SSS. Clusters made at the fourth layer of ART1.5-SSS were three major clusters, two small clusters, and eight clusters, each of which has a unique sample.

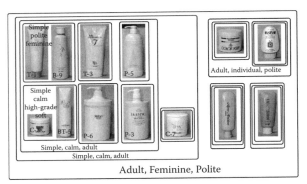

Major cluster 1
white or light color

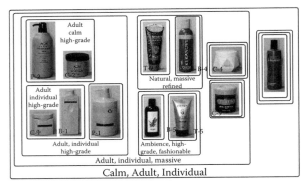

Major cluster 2
deep color
octagonal pot

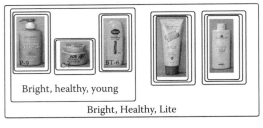

Major cluster 2
bright yellow
or green

Common features: bright blue, yellow green

Small Cluster
horizontal
small product
name

Singleton
clusters
horizontal
small product
name

high-grade, beautiful, simple

FIGURE 6.20
Analysis result of hair treatment with arboART.

First, the major cluster corresponds to Kansei of *adult, feminine,* and *polite,* altogether. Many of the cluster members are colored in white, light pink, or light blue. The shape of the container seems to have no effect. There are several subclusters in this cluster. The upper half of this cluster was formed in the third layer of ART1.5-SSS. This cluster has a large value on *simple, calm,* and *adult.* All members of this cluster are colored in white or a pale color.

The second major cluster corresponds to *calm, adult,* and *individual.* The shape of the container does not have major influences, the same as with cluster 1. The cluster includes (P-2, C-2), (C-2, B-1) and P-1 is characteristic on its dark yellow color. They associated with *adult, individual,* and *high-grade.* The subclusters of T-9 and B-4 are associated with *natural, massive,* and *refined.* Metallic blue and dark green were quite uncommon colors for cosmetic and hair care products at that time. B-5 and T-5 are rather different, but they associate with *ambience, high-grade,* and *fashionable.* C-4 has an ordinary white color, but it has a remarkable octagonal shape.

The third major cluster (P-9, C-8, BT-6, T-8, B-7) has bright blue to green pastel-tone color. This cluster relates *bright, healthy,* and *light.* The members of the first small cluster (BT-2, BT-1, T-6) are common in shape, neutral color, and horizontal small product name. They are *calm, adult,* and *simple.* The second small cluster (T-4, B-8) have a horizontally written name in a small font, and each has a black cap. This cluster corresponds to *high-grade, simple,* and *beautiful.*

From the above results, we confirmed dark yellow containers by Milbon at that time (C-1, B-1, P-1) were associated with *adult, individual, calm,* and *high-grade.* Then, we proposed deep green and blue for another *individual* and *high-grade* color, from T-9 and B-4 of the second major cluster. Those colors were quite novel for hair care products at that time. We also proposed an angular shape for *adult* and *individual* Kansei from C-4 from the same cluster. Finally, we proposed a one-line product logotype in a relatively small font face. These were implied from a small cluster of T-4 and B-8. Consolidating these suggestions, the final product design for Deesee's was determined.

6.3.5 Analysis of Individual Differences

There are many marketing science textbooks that teach segmenting of (potential) consumers. Segmenting itself is an indispensable idea, but preconception about segments sometimes leads the project to fail. Bonnie Goebert, who is an experienced marketing researcher, notes her experiences of gaps between her client companies' preconceived consumer segments and actual consumers of the products (Goebert and Rosenthal, 2001). She warns, "Your intended customer isn't as invested in a product as you are."

When the preconception of different user segments or user groups is difficult, what should be used for classification? We consider that an analysis-based

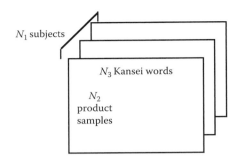

FIGURE 6.21
3-way Kansei evaluation data.

classification of subjects should be developed and utilized. Then, we note our new method for classification of subjects based on *individual clusterings*.

Kansei evaluation data has a three-way structure. *N1* is the number of subjects, *N2* is the number of evaluation samples, and *N3* is the number of Kansei words. Often it was averaged between subjects, then analyzed as a two-way data of *N2* samples × *N3* Kansei words. This research, however, takes up the individual subject's data for the object to analyze. This type of data is called *three-way, three-mode data*.

Although indclus (Arabie et al., 1987) is a well-known method for analyzing individual preference in the framework of cluster analysis, it is difficult to apply to Kansei evaluation data. Indclus is an overlapping clustering technique. This means that an object belongs to multiple clusters. It represents individual differences as weights to clusters. Indclus is favorable when several definitive criteria are assumed in objects of clustering and in subjects. They showed sex difference of subjects on classifying kinship words, defined by sex, by generation, or by degree of relationship.

When we study Kansei for specific products, generally we could not assume such clear criteria (or clusters) for both product and subjects beforehand. The objective of surveying individual difference on Kansei is exploring how many different consumer groups exist and what Kansei is determinative for them. Thus, we will classify subjects by their responding tendency for *N2* × *N3*.

We propose a new method for clustering subjects based on the individual tendencies for Kansei evaluation, for a set of product samples evaluated with a set of Kansei words. First, cluster analysis is done on each subject's evaluation for expressing one's whole tendencies as a hierarchical classification of product samples. Then, the subjects are classified by similarities between each subject's individual clustering. We also show the application of the analysis to our experimental data on milk carton designs and consider the individual differences for the classified subjects (Ishihara et al., 1999).

6.3.5.1 Clustering of Individual Clustering

The basic idea of this approach is clustering subjects by each subject's clustering results. Proximities between subjects are calculated from similarities between individual clustering. Below we describe the three stages of the procedure.

1. Individual clustering. The first stage is clustering evaluation samples (*N2* stimuli) by Kansei words (*N3* attributes) for each subject. Thus, cluster analysis is done for *N1* subjects independently. Virtually any hierarchical clustering method and algorithm are applicable for this stage. In this research, we used arboART for the hierarchical clustering method for individual clustering. It has better clustering performance for large-dimensionality data than traditional methods.

2. Calculating similarities between individual clusterings. In the research of cluster analysis, several studies attempted to compare different clustering results (e.g., Anderberg, 1973; Everitt, 1993). These researchers were focused on finding better clustering methods and plausible classification because there are many formulas for proximity and many clustering algorithms.

   The idea of "clustering of individual clustering" was encouraged by Dubes and Jain (1976). They investigated different algorithms of cluster analysis with test data. The method examined whether a pair of *N2* stimuli was clustered into a same cluster or not. All two-pair combinations of members were checked in two clustering results, then the proximity between the two clustering results was calculated. The method was originally introduced by Anderberg (1973).

   We utilized the methodology of comparing different clustering results to investigate individual differences in general tendencies of Kansei, on samples (*N2* stimuli) × attribute scales (*N3*).

   a. Cutting dendrograms. Usually, hierarchical clustering provides a dendrogram, a tree-like graph that shows the merging process of objects and clusters. The comparison of the clustering results requires cutting the dendrogram at the level of the equal number of clusters. The number of clusters must be considered to represent enough characteristics of the clusters. The number of clusters we used was 1/2 and the number of samples was 1/4 (N2). In our experience, 1/2N2 clusters have a *small* number of members, and those represent rather similar samples. 1/4N2 clusters have *proper abstractions* of similarity structure. If more than two cutting points are used, steps B and C should be *repeated* for each cutting point.

   b. Checking whether a pair of evaluation samples are in the same or different cluster(s). Consider first a single clustering. Assign

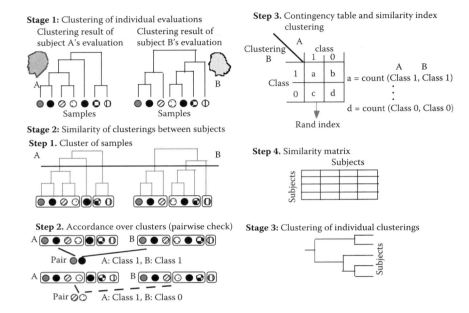

FIGURE 6.22
Procedure of clustering of individual clustering

each of the *N2* (*N2* − 1)/2 pairwise combinations of evaluation sample (stimuli) to one of two classes. Class 0: The samples in the pair belong to different clusters. Class 1: The samples in the pair belong to the same cluster.

c. Calculating similarity of two individual clusterings. When given two clusterings, a contingency table is constructed having the form shown in Figure 6.22. For example, *c* is the number of pattern pairs that are placed in the same cluster by clustering of subject A, but in different clusters by clustering of subject B. Rand's measure of similarity index between the two clusterings is:

$$S(A,B) = \frac{a+d}{a+b+c+d} \qquad (6.25)$$

d. Making a similarity matrix. The similarity index is calculated on pairwise combinations of individual clusterings. The number of combinations is *N1* (*N1* − 1)/2. Calculated values of the similarity index are made into upper triangular elements of an *N1* by *N1* similarity matrix. Diagonal elements are set to zero. If more than two cutting points were used, elements of the matrix are sums of corresponding value.

3. Clustering of individual clusterings. The cluster analysis can be applied when classifying subjects by similarity matrix. In this research, we used UPGMA for the clustering algorithm. Then, clusters of subjects are obtained by their Kansei evaluation on $N2 \times N3$. Each grouped subject has similar individual clustering.

6.3.5.2 Milk Carton Evaluation Experiment and Analysis

6.3.5.2.1 Evaluation Experiment

We conducted an experiment for evaluating 25 milk carton samples on a 5-point SD scale of 69 Kansei words. Subjects were 28 undergraduate students, 25 female and 3 male. They were 20 to 25 years old, and they were not paid for the experiment. The session took two and a half hours with several break times.

Subjects were instructed to take one milk carton sample in hand, view the entire design, then place checkmarks on SD scales.

6.3.5.2.2 Individual Clustering on Each Subject

The ratings on SD scale questionnaires of each subject were analyzed by arboART. Each element of the input vector corresponds to an evaluation of each Kansei word; thus, there were 25 input vectors, which had 69 elements.

For comparison of results, we include the mean of all subject data between subjects. Thus, 29 individual clusterings were used for analysis.

6.3.5.2.3 Clustering of Individual Clustering

Figure 6.23 shows the resulting subject clusters as a dendrogram. We obtained 10 clusters of subjects. The decision method for the suitable number of clusters was seeking large changes of the fusion levels in the dendrogram. Between 8 and 10 clusters had large differences; thus, we decided 10 clusters was representative.

The largest cluster contains the mean of all 28 subjects and 10 other subjects (in the following, we call this cluster an *average cluster*). Other clusters contain from 1 to 4 subject(s) as members. Large-valued elements of the prototype vector of product cluster show high-rated Kansei words on the products. Using the values, we describe the characteristics of some subject clusters.

Figure 6.24 shows the results of individual clustering. Enclosures in the figure should be regarded like a contour map. The outer enclosure represents 6 clusters (1/4 of $N2$) and the inner enclosure represents 12 clusters (1/2 of $N2$), those used for comparing clusterings. Kansei words are high-rated words on the cluster.

Subjects in the average cluster commonly divided all the samples into three or four clusters that had several samples each. Figure 6.24a (upper) shows clustering of the mean of all subjects. Most subject groups were further

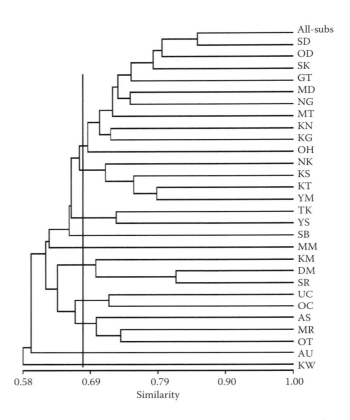

FIGURE 6.23
Dendrogram of clustering of individual clustering.

summarized by design–Kansei relations as three types: (1) The subject groups who made sample clusters similar to average commonly made three clusters; they were abstract designs, cartoon-like illustrations, and logo types that are written from top to bottom with a Japanese writing brush. (2) Subjects KT, YM, KS, and NK were similar to the mean of subjects, although they did not distinguish the logo types, written with brushed calligraphy. (3) Subjects OT, MR, and AS classified designs according to the number of colors. They reacted to cartons using three or more colors and regarded them as *refreshing* and *lively*.

In this way, the analyst can comprehend the difference between subjects in the kinds of design elements to which the subjects respond. Our procedure was easily able to find such individual differences; thus, it may contribute to the research and development of products.

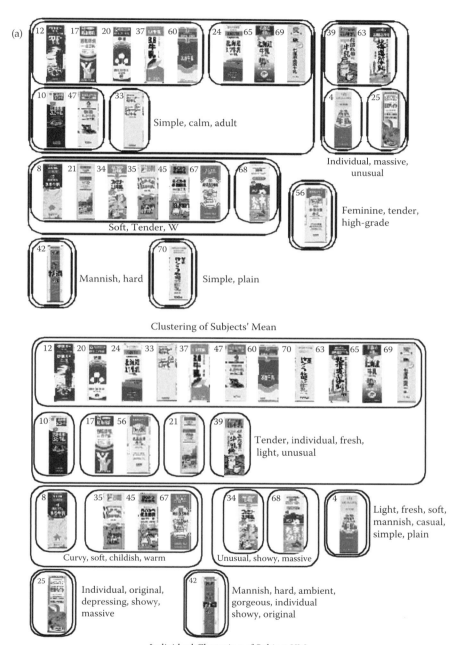

(a)

Simple, calm, adult

Individual, massive, unusual

Feminine, tender, high-grade

Soft, Tender, W

Mannish, hard

Simple, plain

Clustering of Subjects' Mean

Tender, individual, fresh, light, unusual

Light, fresh, soft, mannish, casual, simple, plain

Curvy, soft, childish, warm

Unusual, showy, massive

Individual, original, depressing, showy, massive

Mannish, hard, ambient, gorgeous, individual showy, original

Individual Clustering of Subject YM

FIGURE 6.24
(a) Examples of individual clusterings: Mean of all subjects (upper) and Subject YM (lower).
(b) Examples of individual clusterings: Subject OT.

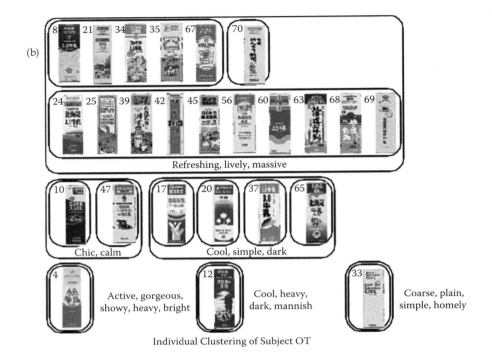

FIGURE 6.24 (continued).

6.4 Genetic Algorithm Model and Kansei/Affective Engineering

6.4.1 Genetic Algorithm Model

Genetic algorithm (GA) is search algorithm or machine learning technology based on natural genetics (Goldberg, 1989). A computer is the complete tool for optimization with the condition that the data of the subject matter can be input in electric format. As we have seen earlier in this book, a problem of the Kansei analyses is converged with optimizations when subjective evaluations and attributes of the candidate are represented as numerical information. This section demonstrates a GA-based searching method to appropriate solutions by means of Kansei evaluation. One concern of KAES is with design of automobile interior spaces. After a brief introduction to the system, the method of optimization by GA was applied to a passenger car interior space design.

The dimensions of the automobile are related to a feeling of a *roomy* and *oppressive* interior. The interior dimensions are a distance or an angle of the vertical or horizontal direction measured from reference points for which

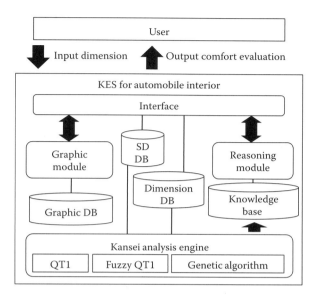

FIGURE 6.25
The components of the automobile interior designing system.

eye point and seating reference point (SgRP) are generally in use. The space is characterized by the physical data of the dimensions. The functional and structural restrictions of the space make it difficult to design the automobile interior. This leads us to the problem of how to obtain the combination of the interior dimensions. This study intends to find the relationship between Kansei and interior space by Kansei engineering, that is to say, the purpose is to design the dimensions to optimize the target product concept represented as the customers' demand.

The automobile interior designing system performs the function of integrating information about interior comfort through dialogue. The system consists of the main modules such as Kansei analysis engine, reasoning, and graphic control modules (Figure 6.25). The graphic module shows sectional drawings of the interior as a guide to recognize the design image and input physical values of the dimensions. The input values are processed at the reasoning module, which correlates the values and specific feelings based on the knowledge base acquired by the Kansei analysis engine. The engine has three kinds of method, QT1, fuzzy QT1, and genetic algorithm. The reasoning module outputs final evaluation of the comfort for the input arrangement of dimensions based on the knowledge base acquired using one of the three methods. The reasoning module also provides the best match dimensions for the set of selected dimensions.

Figure 6.26 shows an example of the sectional drawing illustrated by the graphic module. A user can select several target variables (maximum four items) from the stored dimensions. The system stores 12 sections,

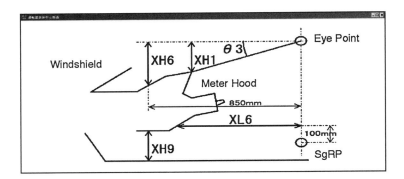

FIGURE 6.26
Install panel section passing through the center of drive's eye point.

53 dimensions in the graphic database. After selecting target items, the user determines the actual measurements for the selected dimensions through scroll input. Then, the system outputs the especially comfortable design, expressed by 100-point scales of *roomy* and *oppression* evaluations from the results of statistical and reasoning modules. At the same time, the reasoning module outputs an optimal range of the selected dimensions for every word with backward Kansei reasoning. We made this interaction system for evaluating interior comfort to support designers at the scene deciding the measurements. In this example we used a mathematical model to correlate input dimensions with actual interior feelings. The main sort to make such KAES is the method to construct the model representing the relationships.

The three engines in the system component were each constructed, respectively, in a different system. The QT1 is used in the most primary system, which was a categorical type of variable representation for the dimensions. Fuzzy QT1 employs a numerical category representation as the fuzzy sets. The QT1 and fuzzy QT1 are both linear regression models. The linear model always contains the problem of multicollinearity. This problem influences reasonability of the backward Kansei engineering method. The GA module is the most advanced method, which represents the reasoning model by a fuzzy decision tree. In the next section, the GA-based method to learn a Kansei model from semantic data is shown.

The mechanism of natural genetics evolves generations through the selection that the individual with high fitness for the environment may survive stochastically. GA is a search algorithm that uses the mechanism of this selection. When dealing with the mechanism on the computer, the individual needs to be defined with the chromosome represented by the symbol string, such as characters or numbers. A new population is reproduced by genetic operations to the individuals. The fitness value to the environment is given by evaluating the chromosome. The individuals with high fitness are selected from the population, and they leave the chromosome to the next generation. The chromosome is composed of two or more genes. The

Item/Category				Chromosome	
Items	Categories	Allele			
Color	Blue	0		[Blue, Round, Small]	: {0,0,0}
	Red	1		[Blue, Round, Big]	: {0,0,1}
				[Blue, Square, Small]	: {0,1,0}
Shape	Round	0		[Blue, Square, Big]	: {0,1,1}
	Square	1		[Red, Round, Small]	: {1,0,0}
				[Red, Round, Big]	: {1,0,1}
Size	Small	0		[Red, Square, Small]	: {1,1,0}
	Big	1		[Red, Square, Big]	: {1,1,1}

FIGURE 6.27
Example of coding to represent items and categories and type of chromosome.

position in which the gene on the chromosome is a locus of a gene and a candidate of a gene is an allelic gene. Each individual is defined by the characteristic of the gene; this characteristic is called a phenotype. To encode is to map from the phenotype into the gene, and to decode is to map from the gene into the phenotype. The allelic gene should be appropriately defined by that variation of the decoded phenotypes that covers the variable space.

Assume that candidates of design elements, which are represented by items and categories, can be mapped by a scheme of chromosome coding. GA can be used for Kansei analysis such as a method using quantification theory or rough set theory, if the elements are represented by certain sets of item/categories. A basic item/category table of the design element and the example of the relation of the chromosome were shown in Figure 6.27. This is a very simple example. Then the definition of the gene that is appropriately expressive of the design space is required in the actual Kansei engineering analysis. For the actual encoding, an indirect method is illustrated using the example of the automobile interior study.

6.4.2 Genetic Algorithm for Automobile Kansei Engineering System

The complexity of human perception is the most problematic factor in Kansei engineering studies. It would be difficult to analyze clear relationships between Kansei data and product design. Several methods of analysis have been proposed that would facilitate a better understanding of Kansei data (Nagamachi, 1995). Multivariate analysis has been the most reliable and available tool. However, even this method will produce erroneous results, usually as a result of nonlinearity. To better understand Kansei in situations where data are limited, we propose extracting the complexity, namely nonlinearity, from the situation.

Interactions between design attributes are the most common cause of nonlinearity. The effects of design combinations are unpredictable and difficult to assess during evaluation because of the occurrence of multiplications, setoffs, or exceptions. Only the use of real data distributions can overcome these difficulties. Then it is important that we discuss Kansei data distributions. The data set used was obtained from a questionnaire-based experiment (using a 5-point SD scale). Design candidates were observed, and those performing the study checked the point.

6.4.2.1 Kansei Evaluation Experiment on Automotive Interior Space

We conducted a Kansei experiment on automotive interior space (Tsuchiya et al., 1996). Forty-one members of an automobile company evaluated 20 passenger cars with 1 liter to 1.5 liter displacements. The number of words for the evaluation was 100. The Kansei words *roomy* and *oppressive* were used to describe the interior feeling. Fifty-three interior dimensions were decided as the measures.

6.4.2.2 Extracting Nonlinear Relations

The automobile interior space showed nonlinear relations between the *roominess* Kansei and the interior dimensions from the distribution check. This type of nonlinearity has been frequently treated as multicollinearity in the field of statistics. It is difficult to estimate statistical correlations between variables. Ishihara advanced a method using local linear regression for splitting design space into linear subspaces to adopt linear regression (Ishihara, 2003; Hastie, 2001). Rough sets theory is expected to be employed to extract effective design conjunctions to represent combination effects. On the other hand, a decision tree can cover whole input space, design elements don't need to be categorical data, and each sample can accept any type of class such as normal sets, numerical value, and mathematical regression model (Tsuchiya et al. 1996, 2005). A classification tree usually has a class. The tree method doesn't require the type of the sample class. The regression model was used at the node in previous research. We may expect that the method is applicable to a large domain of Kansei analysis. The usefulness of the tree method is shown by research using the real product evaluations for extracting important elements from design attributes (Tokumaru, 2002).

6.4.2.3 Kansei Rules by Genetic Algorithm

We attempt to inductively extract the interaction between design elements from the data to better understand the relationship between Kansei data and design elements. Decision trees are popular tools for representing production rules (Quinlan, 1983). There are many ways to construct decision trees, but most are used to minimize quadratic errors. In KAE, the rules used to represent Kansei must be properly structured to fully understand which design elements are interrelated.

The GA-based structure learning is a useful technique for classifying samples. Unfortunately, GA is problematic because it encodes searching

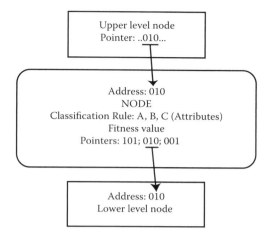

FIGURE 6.28
Indirect encoding to translate tree structure into a chromosome used by genetic algorithm.

problems into chromosomes (Goldberg, 1989), assuming tree structures can be encoded into chromosomes. Because chromosomes tend to decrease in size, trees cannot be directly encoded into chromosomes. In a genetic operator, the population containing the solution is maintained, and this population evolves through successive generations. This is the reason why our method uses indirect encoding (Figure 6.28). Each chromosome corresponds to an attribute, which is a node in the tree. An encoded chromosome has four functioning parts: an address, pointers, classification rules, and fitness values. The address and pointers are described by genotypes (i.e., 0, 1, or don't care), and these form the tree through successive links from the top-level node to the lowest ones. The classification rules place the examples into lower-level nodes according to attribute. Genetic operators maintain a solution population, and the population is allowed to evolve through successive generations. The GA process is shown in Figure 6.29. The tree construction operator generates a tree from the chromosomes in the current population. Then, according to the error rate of the classification, the tree-evaluation operator calculates the constructed tree. The recalculation operator calculates the fitness values for all chromosomes based on the evaluation, and then the genetic operator creates the next population. The most general genetic operators—reproduction, crossover, and mutation—are employed in the process.

6.4.2.4 Resulted Decision Trees

The proposed method is adopted for the interior automobile space to obtain Kansei rules from the evaluation data. Semantic differential data of *roomy* are used for the analysis. Twenty samples of the automobile are classified into five membership functions that correspond to fuzzy classes—class 1

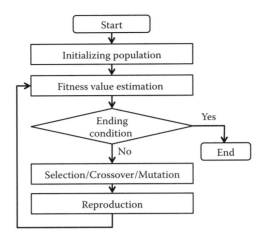

FIGURE 6.29
Flowchart of simple genetic algorithm (SGA).

through class 5—according to the SD values. Sixty-two interior dimensions are categorized into three to five membership functions, which are named A1 to A5. Kansei rules are extracted using our learning method for classifying these training data.

The resulting decision tree for *roomy* is shown in Figure 6.30. The tree is composed of nodes, leaves, and links between nodes. A node means a dimension (e.g., top node is dimension H122). A number of the leaf means a membership function of the fuzzy class, which is proportional to the Kansei evaluation. A link means a membership function of the dimension (A1 means the smallest one, A5 is the largest one). In a decision tree, the level of node represents importance of the attribute. H122 of root node divides the attribute space into three subspaces according to the size of the dimension. A design candidate is classified by the dimension first, which means that H122 is the most effective dimension to classify the samples. The tree also indicates that XH45 is concerned with the classification (dividing attribute space) under the condition of lower H122. Therefore, we may say that our learning method appropriately structured the decision tree according to the data distribution. It also reflects the vertical and horizontal relation between dimensions that dominate in *roomy* for the interior space. Table 6.4 shows a comparison of evaluation values between the real data and prediction by the decision tree. The order of predicted sample evaluation, given by the defuzzify output membership function, is close to the real one. The accuracy of classification was 19/20 in this experiment. This means our method was appropriately able to trace the real data variable. We believe that these results provide enough support for the effectiveness of our method.

The GA-ruled Kansei affective system has been utilized by Isuzu to design passenger car interior design. The designer inputs the expected length and

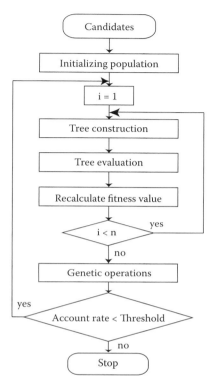

FIGURE 6.30
Flowchart of the tree induction procedure based on indirect genetic algorithm.

angle values into the system, and it responds with results to the designer through the system inference. The system engine estimates the designer's inputted data and sends him/her the recommendation to make a "broader" feeling (*roomy*).

6.5 Virtual Kansei Engineering

6.5.1 Virtual Reality Basis and Kansei Engineering System

KAE is an effective technique for translating human Kansei into the product design element. This technique is utilized in not only Japanese companies but also U.S. and U.K. companies. Especially in Japan, the automobile industry, the home electric industry, and the housing industry have great interest in KAE and are trying to introduce the technique to support the designing process and consumer marketing.

TABLE 6.4

Comparison of Evaluation Values between the Real Data
and Prediction by the Decision Tree

	SD Profile (Real Value)		Decision Tree	
	Sample No.	Rescaled SD Values	Sample No.	Predicted Values
1	Car 7	5.00	Car 7	4.99
2	Car 10	4.46	Car 10	4.45
3	Car 12	4.21	Car 12	4.20
4	Car 6	3.55	Car 6	3.55
5	Car 3	3.41	Car 3	3.40
6	Car 18	2.92	Car 18	2.93
7	Car 15	2.74	Car 15	2.73
8	Car 4	2.64	Car 4	2.59
9	Car 17	2.55	Car 17	2.59
10	Car 5	2.55	Car 5	2.56
11	Car 1	2.46	Car 1	2.38
12	Car 14	2.36	Car 11	2.19
13	Car 11	2.21	Car 19	2.19
14	Car 20	2.21	Car 14	2.15
15	Car 19	2.15	Car 20	2.15
16	Car 2	1.85	Car 13	2.15
17	Car 13	1.79	Car 2	2.01
18	Car 9	1.69	Car 9	1.68
19	Car 8	1.46	Car 8	1.45
20	Car 16	1.00	Car 16	1.01

There are several methods to display the product design on the computer system: text-based information, two-dimensional modeling based on drawing a picture, two-dimensional modeling based on painting a picture, two-dimensional modeling based on photo image (including texture), and three-dimensional modeling. In the previous chapter, we proposed each type of computer system as the Kansei engineering system (KES). On the other hand, virtual reality (VR) is a powerful technology that places the user in a three-dimensional environment that can be directly manipulated, and it is thus useful to introduce VR technique into KES (Zhu 1993, 1994a,b; Enomoto, 1993).

In this chapter, we describe VR-based KES (VR-KES) as the new framework of KES. VR-KES is regarded as the extended system of the three-dimensional modeling-based KES. First, we describe the advantage of introducing VR technology to KES and propose the two types of VR-KES: (1) full-scope VR-KES and (2) desktop VR-KES (walkthrough). Second, we introduce the Kansei analysis support system, KASS, as the designer or system constructor-supporting tool. Then, we focus on a house's system kitchen as the domain for those

VR-KES, show the constructed prototype system, and conclude by evaluating how those two systems investigate the differential of the user's satisfaction.

6.5.2 Virtual Kansei Engineering

6.5.2.1 Integration of VR and Kansei/Affective Engineering

In the traditional KAE framework (Nagamachi 1995; Matsubara 1994), the user (both consumer and designer) can see and experience the candidate design, which the KES infers only as the two-dimensional modeling CG displayed on the CRT. It is difficult to understand the actual total design and concrete depth feeling and usability. It is expected that a lot of changes will occur between the user's desired image and the system-outputted design. Therefore, we introduce VR technology into the KAES framework. In this section, we focus on the *system kitchen* as the system domain. The advantages are as follows:

1. It is easy to experience the location of object unit, the height feeling, and the width feeling.
2. The user can operate the object unit directly (e.g., move the object, open the cabinet door).
3. The user can change the design element detail as easily and at a very low cost (e.g., object unit color, texture).

6.5.2.2 Full-Scope VR and Desktop VR

The VR system is divided into two types of systems: (1) full-scope VR and (2) desktop VR (Thalmann and Thalmann, 1993). The full-scope VR system consists of the main VR system and the stereoscopic display subsystem as the head-mounted display (HMD) or three-dimensional projector, and the sensor subsystem as a data grove, a magnetic sensor or ultrasonic sensor that measures the user's position, and a three-dimensional mouse or space ball that transfers the user's action (Figure 6.31). This system provides the user with a good environment and interface that is very close to a human feeling, but it requires the high-end and large-scale CG computer to process a lot of information. Therefore, many consumers are not able to experience this system conveniently because of its large-size equipment and high cost.

On the other hand, the desktop VR system consists of only a VR main system and a standard display system plus a heavy display subsystem and sensor subsystem (Figure 6.32). This system can be constructed on even a standard personal computer. The consumer cannot experience the stereoscope but can do the walk-through easily. Furthermore, some consumers (i.e., the group consumers) can experience the candidate design at the same time and can discuss whether the proposed design product is good for his own images or not.

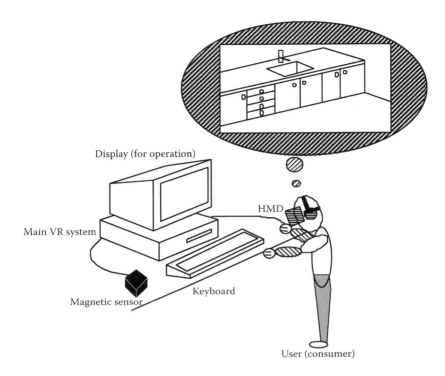

FIGURE 6.31
Outline of full-scope VR system.

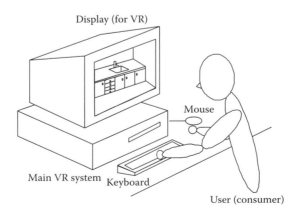

FIGURE 6.32
Outline of desktop VR system.

6.5.3 VR-KES

In this section, we show VR-KES as the new framework of KES: full-scope VR-KES mainly for the designer's use, and desktop VR-KES mainly for the consumer's use. The system configuration is shown in Figure 6.33. The VR-KES consists of (1) the main system, (2) the KES module, and (3) the virtual world construction module. This system is implemented on the Windows PC using the VR platform software (e.g., VRT, World Tool Kit, etc.) and C language, and the input device is a space ball and a mouse.

The main module controls the user ID information and another two submodules. First, the user inputs the facesheet answers, the lifestyle questionnaire answers, and desired image (i.e., Kansei). The main system transfers this information to the KES module. The KES module diagnoses the appropriate candidate design referencing the three databases (lifestyle database, Kansei database, and facesheet database). Then the virtual world construction module gets the diagnosed results and generates the three-dimensional candidate design to combine the three-dimensional object database and texture

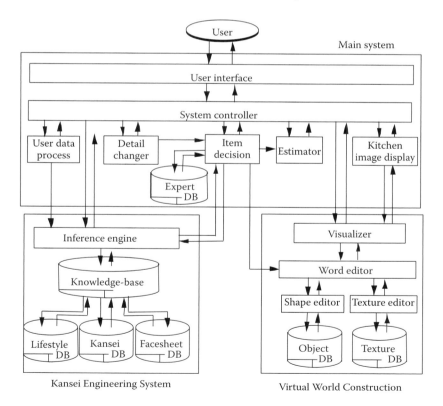

FIGURE 6.33
System configuration of VR-KES.

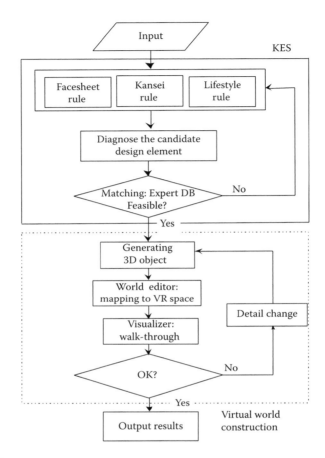

FIGURE 6.34
System flow of VR-KES.

database on the world editor and the visualizer. In this phase, the user can walk through the candidate design and change detail design as he or she likes. Finally, the main module outputs the appropriate design plan, and the user gets much information about the diagnosis results (see Figure 6.34).

6.5.3.1 Kansei Analysis Support System

In the traditional KAE method, it takes a lot of time and manpower to construct the Kansei database using the multivariate analysis method. Moreover it is necessary to do maintenance on the Kansei database about every 5 years to keep the data efficient and to adapt to new trends of product design. As a result, the KES must be updated every 5 years, and the system constructor should be forced to perform the heavy analysis procedure.

KASS has the following function to support the system constructor and to decrease the analysis task based on the concept of the whole in one system:

1. To support the analysis of the raw data from the Kansei experiment
2. To support the building of the database for KES (Kansei database, lifestyle database, facesheet database, etc.)
3. To integrate each database
4. To update and refine the existing database

Especially on the VR-KES, as proposed in the previous section, these functions have the role of providing good customer satisfaction. Figure 6.35 illustrates the KASS concept and the configuration. There are five submodules to support building the database: quantification theory Types I and II, factor analysis, cluster analysis, and dual scaling.

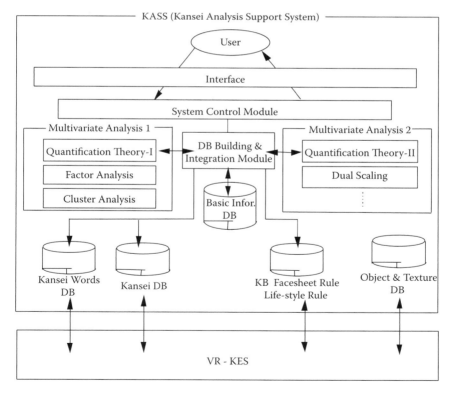

FIGURE 6.35
System configuration of the KASS.

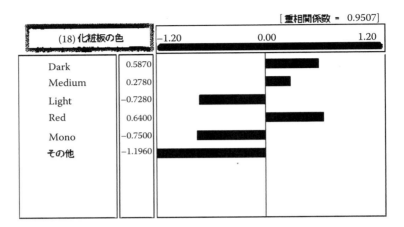

FIGURE 6.36
Output example of QT1.

1. Quantification theory Types I and II. This method is one of the categorical multiple regression analysis method (Hayashi 1976). In the case of Type I, the criterion variable is quantitative and the explanatory variable is qualitative (i.e., categorical parameters). In the Kansei engineering approach, the Kansei estimation value (on the SD scale), which is obtained by Kansei experiment, is treated as the criterion variable, and the design speck, which is expressed as item/category table, is treated as the explanatory variable. The Kansei database is built through this analysis. The output example is shown in Figure 6.36.

2. Factor analysis. In the Kansei engineering approach, factor analysis is used to identify the semantic structure of the Kansei. This result is referenced to determine the *basic Kansei words*, which are the delegates of the adjective word groups, and *general Kansei words*, which include other adjective words. These data are stored as the Kansei words (adjective words) database.

3. Cluster analysis. Cluster analysis is utilized to classify the consumer group. To build the Kansei database, it is required to divide the subjects by age (young/middle/senior), sex (male/female), lifestyle, and so forth. It depends on the designer's or the system constructor's strategy. The output example is shown in Figure 6.37.

4. Dual scaling. The relationship between the design element and the lifestyle, or the design element and the individual attributes (face sheet) are analyzed by the dual scaling method (Nishisato, 1982). The lifestyle database and the facesheet database are constructed using this method and analysis. The output example is shown in Figure 6.38.

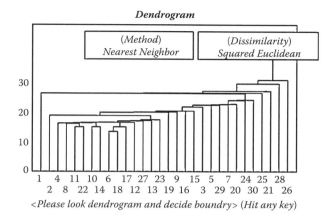

FIGURE 6.37
Output example of cluster analysis.

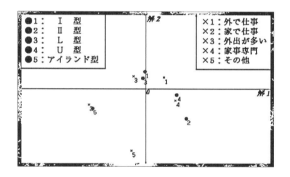

FIGURE 6.38
The output example of dual scaling.

6.5.4 Example

The system's execution examples are shown in Figure 6.39. First the user inputs the personal information (a), then the system generates the first candidate design (b). The user can walk through the virtual space (c) and can attempt the object operations (d). If necessary, the user can change the detail design elements (e).

We give the estimation of the VR-KES considering the difference of display type, walk-through function, moving speed, and individual or group attendance (see Table 6.5) for evaluating 20 items (Table 6.6). The results are shown in Table 6.7, which indicates the mean value of each of the 20 items' subjective estimation about the base system and five other systems. Summarizing the experimental results, we can say the following:

```
INPUT:
        Age: 35
        Height:  156 cm
        Family: 2 generations
                You:: parent
                Persons number::  3
        Last child:  pupil
        Life style:  Life Centered
        Kansei:  Simple
```

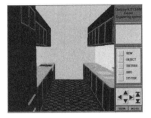

(a) Data input   (b) Inferred first candidate design

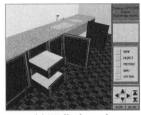   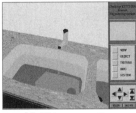   

(c) Walk-through   (d) Object operation   (e) Detail change

FIGURE 6.39
Example of system execution.

TABLE 6.5

Experimental Conditions

	Display	Operational Function	Speed	Subject
base system	CRT	walk-through: s, OP: s	standard	individual: 10
condition 1	[HMD]	walk-through: s, OP: s	standard	individual: 10
condition 2	[TFT]	walk-through: s, OP: s	standard	individual: 10
condition 3	CRT	[nothing]	standard	individual: 10
condition 4	CRT	walk-through: s, OP: s	[low]	individual: 10
condition 5	CRT	walk-through: s, OP: s	standard	[group: 8]

1. Desktop VR-KES (basic system) is an effective tool for the consumer because no complex specific equipment (HMD, etc.) is required.
2. The user often references and uses the walk-through function and object operation function to get decision making support.
3. KES must be supported by at least the walk-through and object operating function.
4. Desktop VR-KES is a good tool to support the group attendance decision making.

TABLE 6.6

Evaluate Items

[General: whole system]		[Decision support function]	
1	pleasant	12	product :: interesting
2	relax	13	product :: useful to buy
3	not :: impatiently	14	easy to understand product
4	not :: fatigue	15	get well-known product
5	not :: worry	16	time reduction to make decision
[Usability]		17	want to buy the product
6	convenient		
7	not :: complex	[General: VR operation]	
8	operation :: good	18	output :: good
9	utility	19	reality :: good
10	operation :: not difficult	20	walk-through :: good
11	operation :: interesting		

TABLE 6.7

Experimental Results

Item	Base System	Cond. 1	Cond. 2	Cond. 3	Cond. 4	Cond. 5
1	6.00	5.90	5.70	4.60**	5.50	5.63
2	5.40	4.20*	5.20	4.90	5.20	5.50
3	5.60	3.30**	4.60	4.70	3.90*	5.38
4	5.40	2.90**	5.10	4.90	4.30	5.38
5	5.70	5.40	5.40	5.20	5.30	6.13
6	6.10	4.70	5.90	4.40**	4.90**	5.63
7	4.50	3.10**	4.40	5.30	4.60	4.63
8	5.80	3.80**	5.80	5.10	4.50**	4.75
9	5.90	4.40**	6.30	4.00**	4.60*	5.38
10	5.30	3.00**	4.60	5.10	3.50**	3.88*
11	5.90	5.90	6.00	4.20**	5.60	6.00
12	5.60	4.90	5.40	4.20**	5.30	5.13
13	6.00	5.20*	6.00	4.50**	5.30	5.88
14	6.00	5.70	6.10	4.20**	5.60	5.75
15	5.50	4.20**	5.30	4.20**	5.40	5.25
16	5.40	4.60*	5.40	3.50**	4.80	5.25
17	4.80	4.20	4.60	3.40**	4.20	4.00
18	5.50	3.30**	5.00	4.30*	5.30	5.13
19	5.70	5.20	5.60	4.10**	5.10	5.13
20	5.90	5.80	5.70	3.90**	5.30	5.13

* $p < 0.05$; ** $p < 0.01$

6.5.5 Conclusions

In this section, we described the VR-KES as the new framework of KES and explained the prototype system using the domain of system kitchen. There are two types of VR-KES: one is the full-scope VR-KES mainly for the designer's use, and the other is the desktop VR-KES mainly for the consumer's use. Both types can be constructed on the same architecture of VR-KES. The differences are the computer graphic processing power and peripheral equipment including the HMD. Second, we introduced KASS as the designer or system constructor supporting tool. Finally, we showed the estimation of the VR-KES and got the results as walk-through and object operating functions, which are the most important factors for KES that support the user's decision making.

References

Anderberg, M.R. (1973). *Cluster analysis for applications*, New York: Academic Press.

Arabie, P., Carroll, J.D., DeSarbo, W. (1987). *Three-way scaling and clustering*. Newbury Park, CA: Sage Publications.

Bayne, C.K., Beauchamp, J.J., Begovich, C.L., Kane, V.E. (1980). Monte Carlo comparisons of selected clustering procedures, *Pattern Recognition*, **12**, 51–62.

Carpenter, G.A., Grossberg, S. (1987). ART2: Self-organization of stable category recognition codes for analog input patterns, *Applied Optics*, **26**, 4919–4930.

Carpenter, G.A., Grossberg, S. (1991). *Pattern recognition by self-organizing neural networks*. Cambridge, MA: MIT Press.

Dubes, R., Jain, A.K. (1976) Clustering techniques: The user's dilemma, *Pattern Recognition*, **8**, 247–260.

Enomoto, N., Nagamachi, M., Nomura, J., Sawada, K. (1993). Virtual kitchen system using Kansei engineering. In: G. Salvendy and M.J. Smith (Eds.), *Human-computer interaction: Software and hardware interface*. Elsevier, Amsterdam, pp. 657–662.

Everitt, B.S. (1993). *Cluster analysis*, 3rd ed., London: Edward Arnold.

Goebert, B., Rosenthal, H. (2001). *Beyond listening: Learning the secret language of focus groups*, John Wiley & Sons.

Goldberg, D.E. (1989). *Genetic algorithm in search, optimization and machine learning*, Boston: Addison-Wesley.

Hamamoto, Y., Uchimura, S., Tomita, S. (1994). Comparison of classifiers in small training sample size situation for pattern recognition, *IEICE Trans. on Information and Systems*, E77-D(3), 1355–1357.

Hastie, T., Tibshirani, R., Friedman, J. (2001). *The elements of statistical learning: Data mining, inference, and prediction*. New York: Springer-Verlag.

Hayashi, C. (1976). *Method of quantification*. Tokyo: Toyokeizai.

Hertz, J., Krogh, A., Palmer, R.G. (1991). Introduction to the theory of neural computation. Redwood City, CA: Addison-Wesley.

Hirasago, K., Jindo, T., Nagamachi, M. (1994). R&D of a design support system for office chairs. *Proc. 12th Triennial Congress of the International Ergonomics Association*, No. 4, Toronto, pp. 127–131.

Imamura, K., Nomura, J., Nagamachi, M. (1994). Virtual space decision support system using Kansei engineering. *Japan–USA Symposium on Flexible Automation-II*, Kobe, pp. 549–555.

Ishihara, S. (2005). Treatment of linearity and non-linearity in Kansei, in Nagamachi, M. (Ed.), *Product development and Kansei*. Tokyo: Kaibundo Publishing.

Ishihara, S., Hatamoto, K., Nagamachi, M., Matsubara,Y. (1993). ART 1.5-SSS for Kansei engineering expert system, *Proc. 1993 Int. Joint Conf. on Neural Networks— Nagoya*, pp. 2512–2515.

Ishihara, S., Ishihara, K., Nagamachi, M. (1994a). AKSYONN: automated Kansei engineering expert system builder by self-organizing neural network, in *Human factors in organizational design and management—IV*. Elsevier: Amsterdam, pp. 485–490.

Ishihara, S., Ishihara, K., Matsubara, Y., Nagamachi, M. (1994b). Self-organizing neural networks in Kansei engineering expert system. *The 11th European Conference on Artificial Intelligence*, Amsterdam, pp. 231–235.

Ishihara, S., Ishihara, K., Nagamachi, M., Matsubara, Y. (1995a). An automatic builder for a Kansei engineering expert system using self-organizing neural networks. *International Journal of Industrial Ergonomics*, **15**(1): 13–24.

Ishihara, S., Ishihara, K., Nagamachi, M., Matsubara, Y. (1995b). arboART: ART-based hierarchical clustering and its application to questionnaire data analysis, *Proc. of 1995 IEEE International Conference on Neural Networks*, Perth, 532–537.

Ishihara, S., Nagamachi, M., Ishihara, K. (1995c). Neural networks Kansei expert system for wrist watch design. In Anzai, Y., Ogawa, K., and Mori, H. (Eds.), *Symbiosis of human and artifact: Future computing and design for human-computer interaction*. Yokohama: Elsevier, pp. 167–172.

Ishihara, S., Ishihara, K., Nagamachi, M., Matsubara, Y. (1997). An analysis of Kansei structure on shoes using self-organizing neural networks, *International Journal of Industrial Ergonomics*, **19** (2), 93–104.

Ishihara, S., Ishihara, K., Nagamachi, M. (1999a). Analyzing method for Kansei engineering data by hierarchical clustering using self-organizing neural networks, *IEICE Trans. on Fundamentals of Electronics, Communications and Computer Sciences* (A), **J82-A** (1), 179–189. (in Japanese).

Ishihara, S., Ishihara, K., Nagamachi, M. (1999b). Analysis of individual differences in Kansei evaluation data based on cluster analysis. *KANSEI Engineering International*, 1(1): 49–58.

Ishihara, S., Ishihara, K., Nagamachi, M. (2000). Kansei analysis and product development of hair treatment. In D. Koradecka, W. Karwowski and B. Das, (Eds.) *Proc. of Ergonomics for Global Quality, Safety and Productivity*, Warsaw, pp. 105–109.

Jindo, T., Hirasago, K., Nagamachi, M., Matsubara, Y., Kagamihara, Y. (1994). A study of Kansei engineering on steering wheel of passenger cars, *Japan–USA Symposium on Flexible Automation—II*, Kobe, pp. 545–548.

Jindo, T., Hirasago, K., Nagamachi, M. (1995). Development of a design support system for office chairs using 3-D graphics. *International Journal of Industrial Ergonomics*, **15**(1): 49–62.

Kashiwagi, K., Matsubara, Y., Nagamachi, M. (1993). The construction of an expert system based on Kansei engineering. *Proceedings of 2nd China–Japan International Symposium on Industrial Management,* International Academic Publishers, Beijing, pp. 545–550.

Kashiwagi, K., Matsubara, Y., Nagamachi, M. (1994). The mechanism of extracting design feature for Kansei engineering. *Human Interface* N&R, **9**(1): 9–16.

Kohonen, T. (1989). *Self-organization and associative memory,* 3rd ed. Berlin: Springer-Verlag.

Levine, D.S., Penz, P.A. (1990). ART 1.5—A simplified adaptive resonance network for classifying low-dimensional analog data, *Proc. of 1990 Int. Joint Conf. on Neural Networks,* Washington, DC, vol. 2, pp. 639–642.

MacQueen, J. (1967). Some methods for classification and analysis of multivariate observations, *Proc. of 5th Berkeley Symp. on Mathematical Statistics and Probability,* vol. 1, pp. 281–297.

Manabe, K., Matsubara, Y., Nagamachi, M. (1994). The rule identification method based on GMM for Kansei engineering design. *Human Interface* N&R, **9**(1): 17–22.

Matsubara, Y., Nagamachi, M. (1994a). An application of image processing technology in Kansei engineering, *Proceedings of 12th Triennial Congress of the International Ergonomics Association,* No. 4, Toronto, pp. 123–126.

Matsubara, Y., Kashiwagi, K., Nagamachi, M. (1994b). The Kansei picture-recognition system and its application to Kansei engineering, *The Institute of Electronics, Information and Communication Engineers,* Technical Report, HC93(72): 47–54.

Matsubara, Y., Nagamachi, M., Jindo, T. (1994c). Kansei engineering as an artificial intelligent system, *Human Factors in Organizational Design and Management-IV,* Elsevier, Amsterdam, pp. 473–478.

Matsubara, Y., Nagamachi, M. (1995). Hybrid Kansei engineering system and design support, *Proceedings of 6th International Conference on Human Computer Interaction,* Yokohama, pp. 161–166.

Michie, D., Spiegelhalter, D.J., Taylor, C.C. (eds.) (1994). *Machine learning, neural and statistical classification,* Chichester: Ellis Horwood.

Nagamachi, M. (1989). *Kansei engineering.* Tokyo: Kaibundo Publishing.

Nagamachi, M. (1993). Kansei engineering and its intelligent implication in product development, *Proceedings of 2nd China-Japan International Symposium on Industrial Management,* International Academic Publishers, Beijing, pp. 539–544.

Nagamachi, M. (1994a). Kansei engineering: A consumer-oriented technology. *Human Factors in Organizational Design and Management-IV,* Elsevier, Amsterdam, pp. 467–472.

Nagamachi, M. (1994b). Implication of Kansei engineering and its application to automotive design consultation. *Proceedings of the 3rd Pan-Pacific Conference on Occupational Ergonomics,* Seoul, pp. 171–175.

Nagamachi, M. (1995). Kansei engineering: A new ergonomic consumer-oriented technology for product development. *International Journal of Industrial Ergonomics,* **15**(1), 3–12.

Nishino, T., Nagamachi, M., Tsuchiya, T., Matsubara, Y., and Cooper, D. (1994). A genetic-based approach to automated design based on Kansei engineering. *Proceedings of the 3rd Pan-Pacific Conference on Occupational Ergonomics,* Seoul, pp. 162–166.

Nishisato, S. (1982). *Quantification method of quaritical data*, Asakura. [In Japanese]

Oja, E. (1982). A simplified neuron model as a principal component analyzer, *Journal of Mathematical Biology*, **15**, 267–273.

Quinlan, J. R. (1983). Learning efficient classification process and their application to chess end games. In Michalski, R. S., Carbonell, J. G., and Mitchell, T. M., (Eds.), *Machine learning: An artificial intelligence approach* (Vol. 1), 463–482. San Mateo, CA: Morgan Kaufmann.

Raudys, S.J., Jain, A.K. (1991). Small sample size effects in statistical pattern recognition: Recommendations for practitioners, *IEEE Transactions on Pattern Analysis and Machine Intelligence*, **13**(3), 252–264.

Sanger, T.D. (1989) Optimal unsupervised learning in a single-layer linear feedforward neural networks, *Neural Networks*, **2**, 459–473.

Sneath, P.H.A., Sokal, R.R. (1973). *Numerical taxonomy*. San Francisco: W.H. Freeman and Co.

Thalmann, N.M., Thalmann, D. (1993). *Virtual worlds and multimedia*. Hoboken, NJ: John Wiley & Sons.

Tokumaru, M., et al. (2002). Decision tree analysis to investigate what has an influence of facility and preference with the product—Making up the rules about database-of-hitting and preference of a golf club. *Journal of Japan Society of Kansei Engineering*, **2**(2): 65–72.

Tsuchiya, T., Matsubara, Y., Nagamachi, M. (1994). A fuzzy rule induction method with genetic algorithm. *Proceedings of the 3rd Pan-Pacific Conference on Occupational Ergonomics*, Seoul, pp. 77–81.

Tsuchiya T., Matsubara Y., Nagamachi, M. (2005). Kansei engineering for design of basic three-dimensional rectangular, *Proc. of the 2005 International Conference on Active Media Technology (AMT 2005)*, pp. 443–448.

Tsuchiya, T., Maeda, T., Matsubara, Y., Nagamachi, M. (1996). A fuzzy rule induction method using genetic algorithm. *International Journal of Industrial Ergonomics*, **18**(2), 135–145.

Zhu, Y., Matsubara, Y., Nagamachi, M. (1993). Development of decision support system of housing design based on Kansei engineering and walk-through technology. *Proceedings of 2nd China-Japan International Symposium on Industrial Management*, International Academic Publishers, Beijing, pp. 523–528.

Zhu, Y., Nagamachi, M., Matsubara, Y. (1994a). Automated housing design expert system using human sensitivity engineering and virtual reality technology. *Proc. of Japan-USA Symposium on Flexible Automation—III*, Kobe, pp. 1211–1214.

Zhu, Y., Nagamachi, M., Matsubara, Y. (1994b). Constructing housing design expert system using human factor and virtual reality technology, *Proc. of the 3rd Pan-Pacific Conference on Occupational Ergonomics*, pp. 262–266.

7

Rough Set Theory and Kansei/Affective Engineering

Tatsuo Nishino

CONTENTS

7.1 What Is Rough Set Theory?

Rough set theory is a mathematical theory founded by Dr. Z. Pawlak (1984) in order to properly deal with vague concepts. The basic idea of rough set theory is that, in contrast to precise concepts, vague concepts cannot be clearly characterized by available information. Kansei words such as *beautiful* and *simple* are vague concepts. The beauty of any product is a vague concept that we cannot always clearly define from its attributes. Therefore, in rough set theory, a vague concept is replaced to the lower and the upper approximation of the vague concept. Lower approximation consists of all objects that *surely* belong to the concept, and upper approximation contains all objects that *possibly* belong to the concept. Two approximations are basic operations in rough set theory (Pawlak, 1998).

Another advantage of rough set theory is that it has some decision rule extraction algorithms to identify the interactive relations between elements in the logical form of *if–then* decision rules that we can easily interpret. Therefore rough set theory provides an effective theory to properly analyze and understand the interactions of elements in complicated relationships between Kansei evaluation and product attributes. As rough set theory can properly deal with vague concepts and it can extract interactions between multiple variables, its application area is rapidly spreading to various fields such as Kansei/affective engineering (KAE), medicine, genetics, databases, computer science, engineering design, marketing research, artificial intelligence, civil engineering, and so on. Rough set theory is very promising especially in the research area involving human judgments and evaluations (Polkowski, Tsumoto, and Lin, 2000).

Nonlinear models such as rough set theory, neural networks (NN), and genetic algorithm (GA) are more effective in KAE applications because KAE deals with vague concepts or parameters such as Kansei words or psychophysical elements and the interactive relations between multiple variables of Kansei evaluations and product elements. In particular, applications of rough set theory to the KAE field are very useful for understanding the complicated relationships between Kansei evaluation and product elements and for supporting design experts to develop more satisfying Kansei products. As extracted decision rules by the rough set rule extraction algorithm are explicitly represented in *if–then* logical forms, design experts can easily interpret these decision rules and can use them in the product design process. In KAE studies so far, some interesting, powerful expert systems such as HULIS system, FAIMS (Fashion Image Expert System), and the hybrid Kansei engineering system have been developed (Nagamachi, 1989, 1995). In constructing KAE expert systems, which are computer-based systems to help design experts develop Kansei products, domain knowledge acquisition is one of bottlenecks. Rough set models make it possible to automatically acquire decision rules from Kansei evaluation data.

Nagamachi (1989, 1995, 2005) has proposed several kinds of KAE methodologies for Kansei product development. Among them, KAE Type I, which mainly uses statistical analysis methods such as factor analysis and quantification theory Type I, has shown great effectiveness for developing Kansei products such as car interiors, entrance doors, fashion, houses, cosmetics, and so on. In Kansei product development, statistical methods are very useful and become necessary tools to find out relationships between Kansei evaluation and product design elements or between Kansei evaluations.

However, these statistical methods are restricted to extracting interactive effects explicitly because these are mainly linear mathematical models. Thus, in order to analyze and understand complicated phenomena such as Kansei, we sometimes must use nonlinear models.

Humans perceive the interactions or gestalt as well as each element of an object, and then recognize or feel it. Accordingly, interactive effects between product elements are a significant factor to be considered in product design. Therefore, we need to develop effective models to explicitly analyze interactive effects between related elements in KAE methodology for product development (Nagamachi, 2006; Mori, 1999; Mori, Tanaka, and Inoue, 2004). In this respect, decision rule extraction algorithms in the rough set theory will help us to extract some interactive effects (decision rules) from complicated data. The rough set approach to KAE methodology for developing Kansei product design is promising because it is effective for analyzing interactions among elements in the form of *if–then*, which can be easily interpreted by product design experts. Though rough set theory appears promising, it should be noted that rough set and statistical models are complementary in KAE practice.

On the other hand, the Kansei evaluation processes are entirely vague in that Kansei evaluation by humans is vague and individual differences in evaluations are not neglected. Rough set theory can properly deal with the vagueness (ambiguity) of Kansei words by approximations of concept. This is one of the reasons why we are much more interested in rough set theory than in other nonlinear models.

Applications of rough set theory have been gradually expanding to KAE research field for the last 10 years. However, many application studies of the rough set model are based on lower approximation of Kansei words, and these studies based on lower approximation are not always effective to extract decision rules reflecting the vagueness of Kansei evaluations. If we can adequately handle the vagueness of Kansei evaluations in the rough set model, we will be able to obtain more significant information on Kansei product design.

In order to reflect appropriately the individual differences among Kansei evaluations in a decision table, our rough set model employs an upper approximation of Kansei word instead of lower approximation. It is a probabilistic rough set model to effectively analyze Kansei evaluation data inspired by the basis of a parameterized version of rough set models, which can explicitly deal with inconsistency in the decision table (Ziarko, 1993; Slezak, 2004, 2005). We identified Kansei words as an indefinable rough set in the framework of the rough set theory, and we have developed a probabilistic rough set model and computer program that represents probabilistically the Kansei or human evaluation data with much ambiguity (Nishino, 2005; Nishino, Nagamachi, and Tanaka, 2005, 2006).

Further, we have been examining the effectiveness of our model in some Kansei product design applications such as children's shoes, coffee taste, comprehensive ball pens, and car floor mats (Nishino, Nagamachi, and Sakawa, 2006; Hirata, Nishino, and Nagamachi, 2007). Moreover, we have proposed a rough set approach using a multilevel framework in KAE methodology that consists of four levels to extract decision rules for Kansei product development: product development goal, customer Kansei needs, developmental concepts, and product design attributes (Nishino and Nagamachi, 2007a, 2007b, 2009; Nishino, 2009a, 2009b).

At the present research stage, applications of the proposed rough set model to Kansei product designs have been very promising.

7.2 Theoretical Structure of Rough Set Theory

In this section, we introduce some basics of rough set theory (Pawlak, 1998) and the extended rough set model that we have been developing with the practical aim of applying it to Kansei product design.

7.2.1 Basics of Rough Set Theory

7.2.1.1 Decision Table

For simplicity, we first describe the basic elements of rough set theory by means of an example. Data are often presented as a decision table, with columns labeled by attributes and rows by objects, whereas entries of the tables are attributes values. Attributes are sometimes divided into two classes, condition and decision attributes.

For example, Table 7.1 contains information about objects (products) evaluated by the Kansei word *beautiful*. The attributes *color, size,* and *shape* can be considered as condition attributes, whereas the attribute *beautiful* is a

TABLE 7.1

Simple Decision Table

Product	Color	Size	Shape	Kansei (beautiful)
p1	black	large	triangle	yes
p2	red	small	triangle	yes
p3	red	large	circle	yes
p4	black	large	square	no
p5	red	small	triangle	no
p6	black	large	circle	yes

decision attribute. Such data forms are known as decision tables. Each row of a decision table shows a condition–decision relationship, which specifies the decisions (evaluations) occurring when the conditions are satisfied. For example, the conditions (*color, black*), (*size, large*) and (*shape, triangle*) of product *p1* determine uniquely the decision (*beautiful, yes*). Notice that data corresponding to products *p2* and *p5* in Table 7.1 are the same conditions but different decisions. Decision tables with such data are called inconsistent (nondeterministic, conflicting, possible); otherwise, the tables are referred to as consistent (certain, sure, deterministic, nonconflicting).

7.2.1.2 Approximations of Vague Concepts

In Table 7.1, product *p2* is evaluated as a beautiful product, whereas product *p5* is not evaluated as a beautiful product, and they are indiscernible (the same) with respect to the conditional attributes *color, size,* and *shape*; thus, Kansei *beautiful* cannot be characterized in terms of *color, size,* and *shape*. Therefore, products *p2* and *p5* are in a boundary region, which cannot be properly classified by using the available knowledge. The remaining products *p1, p3,* and *p6* can be classified as *beautiful,* and product *p4* as certainly *not beautiful*.

Thus the lower approximation of Kansei word *beautiful* is the set {*p1, p3, p6*} and the upper approximation of this concept is the set {*p1, p2, p3, p5, p6*}. Similarly, as product *p4* is certainly *not beautiful,* the lower approximation of the concept *not beautiful* is the set {*p4*} and the upper approximation is the set {*p2, p4, p5*}.

This means that upper approximation of the concept includes the boundary region, but the lower approximation excludes the boundary region. Using these approximated regions and the decision functions described in the next section, we can extract decision rules in the form of *if–then*. Figure 7.1 illustrates a relation between concept approximation and decision rule extraction using the rough set model.

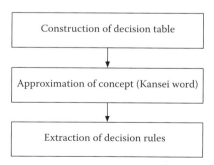

FIGURE 7.1
Rule extraction by rough set.

7.2.2 Extended Rough Set Theory

7.2.2.1 Decision Table for Kansei Evaluation Data

Now we introduce an extended rough set model that we have been developing for use in Kansei engineering applications (Nishino, 2005, 2009a; Nishino, Nagamachi, and Tanaka, 2005, 2006). Table 7.2 shows simple representative Kansei experimental evaluation data using Kansei words. Object set U is the set of evaluation events $U = \{x_{11}, ..., x_{54}\}$. An evaluation event of jth evaluator to ith product is denoted as x_{ji}. There are four products and five human evaluators. Each attribute of A has a domain of its design attribute values, $Va_1 = \{0,1\}$, $Va_2 = \{0,1,2\}$, and $Va_3 = \{0,1,2\}$. Product attribute set A can be {color, function, and shape}, and attribute value set in color attribute can be {black (0), red (1)}. Partition of U by attribute set A constitutes the set of equivalence class. For example, evaluation events from x_{11} to x_{51} constitute an equivalence class $E_1 = \{x_{11}, ..., x_{51}\}$, which has the same attribute value with respect to every conditional attribute set A. Similarly, evaluation events from x_{12} to x_{52} are an equivalence class. Thus, we can obtain four equivalence

TABLE 7.2

An Example of Kansei Evaluation Data

Product (E)	Event (U)	a_1	a_2	a_3	Simple (d_1)	Beautiful (d_2)
E_1	x_{11}	0	1	1	0	1
	x_{21}	0	1	1	0	0
	x_{31}	0	1	1	0	0
	x_{41}	0	1	1	1	1
	x_{51}	0	1	1	1	0
E_2	x_{12}	1	0	2	1	0
	x_{22}	1	0	2	1	0
	x_{32}	1	0	2	1	1
	x_{42}	1	0	2	0	1
	x_{52}	1	0	2	1	1
E_3	x_{13}	0	2	0	1	1
	x_{23}	0	2	0	1	1
	x_{33}	0	2	0	1	1
	x_{43}	0	2	0	0	0
	x_{53}	0	2	0	0	0
E_4	x_{14}	1	1	1	0	0
	x_{24}	1	1	1	0	1
	x_{34}	1	1	1	0	1
	x_{44}	1	1	1	0	1
	x_{54}	1	1	1	1	1

classes E_1, E_2, E_3, and E_4 in the table. The set of equivalence class is denoted as $E = \{E_1, E_2, E_3, E_4\}$. Human evaluation d also has a domain of its evaluation values, $V_{d1} = \{0,1\}$, $V_{d2} = \{0,1\}$, which may be *not simple* (0) and *simple* (1) or *not beautiful* (0) and *beautiful* (1). The concept *simple* is defined as $X_1 = \{x \mid d_1(x) = 1\}$. In Table 7.2, we can define elementary concepts such that $X_1 = \{x \mid d_1(x) = 1\}$, $\neg X_1 = \{x \mid d_1(x) = 0\}$, $X_2 = \{x \mid d_2(x) = 1\}$, $\neg X_2 = \{x \mid d_2(x) = 0\}$.

Further, in order to deal with complex Kansei words in Kansei engineering, we can define the combined concept Y of elementary concepts X_i as

$$Y = X_1 \circ X_2 \circ, ..., \circ X_m \tag{7.1}$$

where the operator is a logical operator ∧*(and)* or ∨*(or)*.

For example, a combined concept (Y) of *simple* (X_1) and *beautiful* (X_2) is obtained by $Y_1 = X_1 \wedge X_2$, $Y_2 = X_1 \vee X_2$, $Y_3 = X_1 \wedge \neg X_2$ and so on. The rough set model extracts decision rules for these combined concepts.

Rough set theory can properly deal with inconsistent tables where conditional attribute values are the same but decision attribute values are different. For example, in response to the equivalence (product) class E_1, two people evaluated *simple*, but the other three people evaluated *not simple*. Evaluation is also not consistent for the other products. The basic idea of rough set theory is to deal with such vague concepts properly. Even if a lot of design attributes are considered in decision tables, vagueness often appears in data tables that include human evaluations or judgments. Generally, we can formalize Kansei evaluation data obtained by KAE method as a nondeterministic decision table as shown in Table 7.2.

7.2.2.2 Kansei Approximation

The lower approximation D_* and upper approximation D^* of concept $D = \{x : d = 1\}$ are defined using equivalence class E_i partitioned by design attribute sets A as follows:

$$D_* = \{E_i \mid E_i \subseteq D\} \tag{7.2}$$

$$D^* = \{E_i \mid E_i \cap D \neq \phi\} \tag{7.3}$$

The set $BN = D^* - D_*$ is a boundary region of D. In this table, we noticed that we cannot obtain the lower and upper approximation of concept *simple* D_1 because $D_{1*} = \phi$ and $D_1^* = U$. Such concepts belong to totally undefinable concepts in four categories of vagueness of the rough set theory (Pawlak, 1998). Therefore, the concept approximation from a decision table of Kansei evaluation data obtained by Kansei words is generally a difficult task.

On the other hand, we can extract decision rules by constructing a consistent (deterministic) decision table using average scores among evaluators. Regrettably, the decision rules extracted by the average method cannot properly reflect the vagueness of Kansei words: this is one of the very interesting and serious properties of Kansei words.

Generalized rough set models such as the variable precision rough set (VPRS) and the probabilistic rough set (BRS) are proposed using a consistent degree or precision parameter, not inclusion relationships between sets. Based on these generalized rough set models, we have developed a rough set model to properly treat the vague properties of Kansei words and to effectively use in KAE.

The Kansei evaluation data shown in Table 7.2 include at least two important probabilistic aspects. One is the probability $P(Y|E_i)$ of decisions dependent on the attributes of product E_i and the other is the prior probability $P(Y)$ of decision class Y. Such probabilities are experientially acceptable in human evaluation data. These probabilities are well known as the conditional and prior probability, $P(Y|E_i)$ and $P(Y)$, respectively.

We define the information gain of E_i with $P(Y|E_i) \geq P(Y)$ denoted as

$$G_{pos}(Y, E_i) = 1 - P(Y)/P(Y|E_i) \tag{7.4}$$

which means that the greater the conditional probability relative to prior probability, the greater the information gain.

We define the positive region by using the information gain with the parameter β as

$$POS^\beta(Y) = \cup\{E_i \mid G_{pos}(Y \mid E_i) \geq \beta\}$$
$$= \cup\{E_i \mid P(Y \mid E_i) \geq P(Y)/(1-\beta)\} \tag{7.5}$$

which means the region that E_i possibly belongs to is Y with β.

$$NEG^\beta(Y) = \cup\{E_i \mid P(Y \mid E_i) \leq (P(Y)-\beta)/(1-\beta)\} \tag{7.6}$$

which means the region that E_i does not possibly belong to is Y with β.

$$BND^\beta(Y) = \cup\{E_i \mid P(Y \mid E_i) \in ((P(Y)-\beta)/(1-\beta), P(Y)/(1-\beta))\} \tag{7.7}$$

which means the region that E_i does not belong to is neither Y nor $\neg Y$ with β. As the value of β increases, the positive and negative regions decrease, and boundary region increases. Furthermore, as the value of β increases, the information associated with D_j is strongly relevant to E_i.

In addition, we can define:

$$UPP^\beta(Y) = POS^\beta(Y) \cup BND^\beta(Y)$$

$$= \cup\{E_i \mid P(Y \mid E_i) \geq (P(Y) - \beta) / (1 - \beta)\}$$

(7.8)

which means the possible region being larger than POS^β (Y).

We call POS^β (Y) the β-lower (upper-lower) positive approximation of Y, and UPP^β (Y) the β-upper (upper-upper) positive approximation of Y.

In order to compute decision rules for these approximated decision classes, we will use a discernibility matrix, which means $m \times n$ matrix, rows of which are product set E_i $(i = 1, \ldots, m)$ belonging to approximated decision class (for example POS class), and columns are product set E_j $(j = 1, \ldots, n)$ belonging to the other approximated class (for example, NEG and BND classes). In this case, the decision rules of POS class will be extracted.

7.2.2.3 Extraction of Decision Rules

Discernible entry elements in the matrix are as follows:

$$\delta(E_i, E_j) = \{(a, a(E_i)) \in A : a(E_i) \neq a(E_j)\}$$

(7.9)

The entry $\delta(E_i, E_j)$ is the set of all attribute–attribute value pairs that discern product set E_i and E_j. Thus, the discernibility matrix image with respect to POS class is shown in Table 7.3.

We can derive decision rules by decision functions from approximation regions as follows:

$$f^{pos}(A) = \bigvee_{E_i} \wedge \bigvee_{E_j} \delta(E_i, E_j)$$

(7.10)

where $\vee\delta(E_i, E_j)$ $\vee$ and $\wedge$ denote Boolean *or* and *and*, respectively.

For example, by constructing the decision function from Table 7.1 we can obtain decision rules like "if (*color, red*) and (*size, large*), then (*beautiful, yes*)."

TABLE 7.3

An Example of Discernibility Matrix

	Neg	*Bnd*	
	E_4	E_2	E_5
Pos	E_1 E_3 E_6	δ (E_i, E_j)	

7.2.3 Evaluation Measures of Decision Rules

The extracted decision rule is denoted as *if* Φ *then* Ψ. Then, we define the following measures for decision rules:

$$Supp(\Phi, \Psi) = \|\Phi \wedge \Psi\| \tag{7.11}$$

where $\|\cdot\|$ is cardinality. This measure indicates the number of events satisfied with decision rules.

$$Str(\Phi, \Psi) = \frac{Supp(\Phi, \Psi)}{\|U\|} \tag{7.12}$$

$$Cer(\Phi, \Psi) = \frac{\|\Phi \wedge \Psi\|}{\|\Phi\|} \tag{7.13}$$

$$Cov(\Phi, \Psi) = \frac{\|\Phi \wedge \Psi\|}{\|\Psi\|} \tag{7.14}$$

Certainty $Cer(\Phi, \Psi)$ indicates the extent to which if its conditions are satisfied, a decision can be derived. Inversely, *coverage* $Cov(\Phi, \Psi)$ indicates the extent to which if its decision is satisfied, a condition can be derived.

The above four measures are affected by the frequency of attribute value in the data table. If product samples are assumed as random ones, these measures will reflect properly the set of events. However, if not, we had better use the following effect measure in order to measure the real effect of decision rules.

$$Eff(\Phi, \Psi) = \frac{\|\Phi \wedge \Psi\| \cdot \|U\|}{\|\Phi\| \cdot \|\Psi\|} \tag{7.15}$$

If the value is larger than 1, the effect of decision rule is significant. The measure generally indicates effective combination of condition parts. By combining support and effect measures, we may use the following measure for an example:

$$Com(\Phi, \Psi) = Eff(\Phi, \Psi) + c \cdot Str(\Phi, \Psi) \tag{7.16}$$

Using different rule evaluation measures will make it possible to extract different types of decision rule sets.

7.2.4 Multilevel Framework for Kansei Design Rule Extraction

In KAE for product development, it is effective to link each of the other three processes for discovering the customer's need for the product, searching developmental concepts for a new product satisfying customer Kansei needs, and finally discovering the product attribute values to realize developmental concepts. Conducting three processes separately may not be effective for promoting the new product development. Therefore, we proposed the multilevel framework for Kansei design rule extraction as shown in Figure 7.2, which consists of four analysis process levels, and decision rules between levels are extracted using the rough set model: goal, customer Kansei needs, developmental concept, and design attribute (Nishino and Nagamachi, 2007a, 2007b, 2009).

Goal level indicates decisive level in regard to products as shown by decisive words such as *want to buy* and *attractive*, which link the customer's decision to purchase the product. Kansei need level means deeper Kansei needs of customers, which cannot be sufficiently represented in terms of Kansei words. Concept level indicates the language representation of the customer Kansei needs, which can be used as product development concepts. Design attribute level means product design values satisfying the customer needs and the development concept.

An advantage of the multilevel framework is to provide a method to analyze sequentially and systematically customer Kansei needs and development concept as well as design elements. In the framework, the rough set model is used to extract three types of decision rules among levels as shown in Figure 7.2: goal–needs rule, needs–concept rule, and concept–design attribute rule. Decision rule extractions are conducted top down from goal to design attribute. For example, it extracts the combination between principal component elements satisfying the customer-purchasing motive (decisive Kansei words) in principal component space: deeper customer Kansei needs.

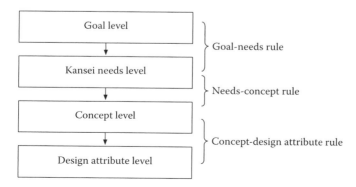

FIGURE 7.2
Multilevel decision rule extractions.

It is possible to extract different types of decision rule sets at each step using one of three different rule evaluation measures: support (S), effect (E), and combined (C). For example, rule extraction using S-measure at all three steps, S-S-S strategy, extracts decision rule with more customer supports at each extraction step. In order to search rather ordinal design attributes, this S-measure will be effective. E-E-S strategy extracts more unique and effective product design rules for customers. Therefore, we have to select a good sequence of rule evaluation measures according to a company product development policy and product position in the market place.

7.3 An Application of Rough Set Theory

In this section, we will describe an example to show the effectiveness of our rough set model in Kansei product design. The purpose of the research is to find out important design attribute values and their combinations, and to generate the possibilities of new ball pen design for young female customers from the viewpoint of Kansei product design (Nishino and Nagamachi, 2007a, 2009).

7.3.1 Kansei Evaluation Experiment

We arranged 27 ball pens from different companies for an evaluation experiment. Twenty-four female students evaluated each ball pen while seeing and touching it, using 5-point semantic differential (SD) scale measurements consisting of 40 Kansei words and one decisive (goal) word, *attractive*. Kansei words are words representing affective and operational aspects of products such as *young, advanced, high-grade*, and *easy to use*. Decisive words are words directly linking customers purchasing, such as *want to buy, attractive*, and *like*. In this case, the word *attractive* is used. Design elements of ball pens were divided into 22 attributes and 66 attribute values including functional, and operational and decorative attributes such as body color, weight, material of body, and so on. We used a dichotomy of Kansei word evaluation scores for probability computation for the rough set model.

7.3.2 Principal Component Analysis

We computed principal component analysis of 40 Kansei words except the decisive word *attractive* as shown in Table 7.4. Using accumulated contribution criteria of 85%, we could obtain five principal components of 40 Kansei words: *advanced-practical, beauty-handiness, young-high grade, friendly-original*, and *simple-functional* components. The advanced-practical component includes Kansei words such as *advanced, playful, good buy, good design, stylish, want to buy*, and so on. The beauty-handiness component includes Kansei words such as *easy*

TABLE 7.4

Principal Components

Principal Component	Eigenvalue	Contribution (%)	Accumulated Contribution (%)
Advanced-Practical	18.45	45.00	45.00
Beauty-Handiness	7.95	19.40	64.40
Young-High grade	4.29	10.46	74.86
Friendly-Original	2.81	6.86	81.72
Simple-Functional	1.53	3.73	85.45

to use, easy to grip, not tired, easy operation, and so on. The young-high grade component includes *young, colorful, good color harmony, elegant, high-grade, high price,* and so on. The friendly-original component includes *slim, friendly, original, easy to find,* and so on. The simple-functional component includes *simple, curved line, female-oriented, multifunctional,* and so on.

From the result, one can easily see that young females want ball pens characterized by advanced and beauty components because the contribution ratios of these components are higher than the other components. Although we can estimate such general directions of customer needs from the result of principal components, for designing a more satisfying product, more accurate Kansei needs the directions of young females to be estimated. By using the rough set model, we can obtain some potential directions of customer Kansei needs in principal component space.

7.3.3 Customer Kansei Needs

By constructing a decision table consisting of the category of the principal component scores of Kansei words and the conditional probability of the decisive word *attractive,* we can obtain the combinations of component axes of Kansei words satisfying the decisive word *attractive* in principal component space. Decisive words such as *want to buy, attractive,* and *like* are direct representations of total motivation to the product, whereas Kansei words are indirectly linked to total motivation to the product. We can assume that combinations of principal component categories satisfying with a decisive word will represent customer Kansei needs. Knowledge of Kansei words and decisive words will help the product engineer or designer get a good idea for the product design from the viewpoint of customer needs.

In order to compare the results of the rough set method, the results of multiple regression analysis often used in marketing research (Kanda, 2000) are shown in Table 7.5. An objective variable is the decisive word *attractive,* and the explanation variables are five principal components. From the result, we can suppose that many young female customers want *advanced* ball pens. However, from the result of multiple regression analysis, it is difficult to estimate the combinations of principal components to explain *attractive.*

TABLE 7.5

Multiple Regression Analysis for *Attractive* Ball Pen

	Decision coefficient (R^2)	0.90296
	Multiple correlation (R)	0.95024
Principal Component	**Regression Coefficient**	**STD Coefficient**
Advanced-Practical	0.088	0.913
Beauty-Handiness	0.034	0.233
Young- High grade	0.013	0.062
Friendly-Original	0.024	0.095
Simple-Functional	0.012	0.036

Table 7.6 shows a decision table for rough set analysis including component score category as conditional attributes and the conditional probability of the goal of Kansei *attractive* as a decision attribute. The symbols *A* to *E* in the table indicate principal components. A principal component score is divided into two categories using the sign of the component score. For example, A1 and A2 indicate plus and minus score categories of the advanced-practical component axis, advanced and practical, respectively. Each row indicates the relation between the category of five principal components and probability of Kansei word *attractive* regarding a product. For example, product 8 has a higher probability of *attractive* conditioned by component categories of A1, B1, C1, D2, and E1. By applying the rough set model to the table, we can obtain the combinations of component categories satisfying the decisive word *attractive* in principal component space.

Tables 7.7 and 7.8 show only the top four decision rules extracted by the rough set model with S-measure and E-measure, respectively. Decision rules extracted with S-measure derive most general combinations of the component, while decision rules extraction with E-measure derive more specific ones. Extracted principal component directions with S-measure indicate that there are four clusters of customer Kansei needs and these rules have relatively simple short conditions. For example, the cluster with the most dominant needs is *advanced*. The second cluster is *handiness, young*, and *original*. Thus, as S-measure extracts more supportive rules, we can assume that these Kansei needs will be supported by many young females.

Clusters with E-measure as shown in Table 7.8 indicate effective-needs clusters for product development because they have higher effective values. Clusters by effect (E) measure indicate effective clusters for product development because they have rather higher effective values. These rules have relatively long specific conditions. More dominant clusters are *simple, advanced*, and *beauty*; and *simple, advanced*, and *young*. Extracted rules with E-measure are more effective strong rules. This suggests that there are some small clusters of customers with unique strong Kansei needs, and new ball pens satisfying these Kansei needs will be able to much improve the attractiveness to the customers.

TABLE 7.6

Decision Table for Customer Kansei Needs Identification

Product	A	B	C	D	E	No. of Subjects	Attractive
1	A1	B1	C2	D2	E1	24	0.500
2	A1	B1	C1	D1	E2	24	0.500
3	A1	B1	C2	D1	E2	24	0.542
4	A2	B1	C1	D2	E2	24	0.167
5	A1	B1	C1	D1	E2	24	0.417
6	A2	B1	C1	D2	E1	24	0.125
7	A1	B2	C1	D1	E2	24	0.417
8	A1	B1	C1	D2	E1	24	0.625
9	A1	B2	C2	D1	E2	24	0.417
10	A1	B2	C1	D2	E2	24	0.583
11	A2	B2	C2	D1	E2	24	0.458
12	A1	B2	C1	D2	E2	24	0.583
13	A2	B2	C2	D2	E2	24	0.125
14	A2	B1	C2	D2	E2	24	0.417
15	A1	B2	C1	D2	E2	24	0.458
16	A2	B2	C2	D2	E1	24	0.250
17	A2	B2	C1	D2	E1	24	0.375
18	A2	B2	C1	D1	E1	24	0.250
19	A2	B1	C2	D2	E1	24	0.417
20	A1	B2	C1	D1	E1	24	0.500
21	A1	B1	C1	D1	E1	24	0.583
22	A2	B2	C2	D2	E1	24	0.417
23	A1	B2	C1	D2	E2	24	0.458
24	A1	B2	C2	D1	E2	24	0.417
25	A2	B2	C1	D2	E1	24	0.375
26	A2	B1	C2	D1	E1	24	0.333
27	A1	B2	C2	D1	E1	24	0.375

TABLE 7.7

Customer Needs Directions by Support Measure (S)

Customer Needs Direction	Supp	Cer	Eff
Advanced	177	0.492	1.198
Friendly and Functional	76	0.452	1.102
Handiness and Young and Original	68	0.472	1.150
Handiness and Young and Functional	60	0.500	1.218

TABLE 7.8

Customer Needs Directions by Effect Measure (E)

Customer Needs Direction	*Supp*	*Cer*	*Eff*
Simple and Advanced and Beauty	41	0.569	1.387
Simple and Advanced and Young	41	0.569	1.387
Advanced and Original	77	0.535	1.303
Handiness and Original and Young and Functional	50	0.521	1.269

7.3.4 Developmental Concept Rule

We found two clusters of young female Kansei needs of *advanced* extracted with S-measure, and *simple, advanced*, and *young* with E-measure. In a similar way, we computed the combinations of Kansei words satisfying these customer Kansei needs at the second step.

For the customer Kansei needs cluster of *advanced*, single words such as *advanced, want to buy*, and so on were extracted with S-measure as developmental concepts. This indicates that an advanced ball pen is attractive for a lot of young females. On the contrary, extracted rules for customer needs clusters of *simple, advanced*, and *young* were very unique development concepts such as *playful, not multifunctional, not high-grade, good buy*, and so on. Generally speaking, these rules suggest that *playful* and *good buy* ball pens will be more attractive for some young females. Which is a better developmental concept—advanced, or playful and good buy—is dependent on the product development policy of the company. For comparing product design rules derived from these different concepts, we computed each product design rule satisfying *advanced*, and *playful*, and *good buy*, respectively.

7.3.5 Product Design Rule

In the constructed decision table for rough set analysis, condition attributes are 22 design attributes of ball pen such as body color, weight, and material of body, and 66 attribute values are used. For example, the color attribute has nine design values: red, pink, orange, brown, green, blue, white, black, and transparent. The decision attribute is a developmental concept—advanced, or playful and good buy—which is represented as a conditional probability in the decision table.

More than 200 decision rules are extracted from each table. Table 7.9 shows the *if* part of the extracted design rules for the concept *advanced*, which are ordered by descending order of S-measure score. One can easily see that the three-color mechanics in a ball pen is an important design attribute value for the advanced concept and this design value interacts with other mechanical design values such as plastics material, separate top joint, and rubber grip. These design rules indicate the number of mechanics of significant design attributes satisfying many young females who have Kansei needs *advanced*.

TABLE 7.9

Decision Rules for *Advanced* Concept

IF Part
Material (plastics) and No. of mechanics (three colors)
To join (separate) and No. mechanics (three colors)
Automatic pencil (no) and No. mechanics (three colors)
Grip quality (rubber) and No. of mechanics (three colors)
…
Color (brown)

TABLE 7.10

Decision Rules for *Playful* and *Good Buy* Concept

IF Part
Color harmony (two colors) and No. of mechanics (three colors)
Color harmony (two colors) and Clear (no)
Color harmony (two colors) and Weight (16g +)
. . . .
Color (blue)

Table 7.10 shows the *if* part of the extracted design rules for the concept *playful* and *good buy*. One can easily see that color harmony of two colors is an important design value for the playful and good buy concept and this value also interacts with other design values such as three-color mechanics, not clear, and weight more than 16 grams. These rules represent more attractive product design specifications for some young females who have Kansei needs of *advanced*, *young*, and *simple*. These design rules indicate that color harmony is more attractive for some customers than the number of mechanics.

These results suggest that those development concepts are related to combinations of design values. It is difficult for design experts to find them out, and combinations of design values should be considered in Kansei product design. In this point, the rough set model provides a tool to find out the interactions between design attribute values in Kansei product design. From extracted product design rules, product design experts will be able to get significant information on an attractive product, which it is difficult to acquire experientially. The product designer/developer also can obtain new ideas on product design by checking inconsistent decision rules and negative ones.

Thus, we could find the developmental concepts and its design specifications from two paths for extracting product design elements for developing attractive ball pens. This suggests that there is not one best rule set of design values for attractive product design and whether a product is attractive is dependent on customer Kansei needs clusters. The multilevel framework based on the rough set model makes it possible to link customer Kansei needs to attractive design attribute values.

7.3.6 Rule-Based Design System

In order to effectively use a lot of extracted product design rules in Kansei product design, we are trying to construct a rule-based design system that includes the extracted design rules by the rough set model. It automatically searches design specifications using the design algorithm that applies decision rules sequentially to a design table in descending order of the measure score of design rules and determines design values in all design attributes. While only consistent decision rules with current states of the design table are applied to a design table, conflicting ones are skipped. The algorithm is very simple, but we can use another more complicated algorithm and get different designs. Table 7.11 shows the product design value sets suggested

TABLE 7.11

Design Values Derived from *Attractive*

Goal	Attractive Product	
Customer needs (component level)	Advanced	Advanced, young, and simple
Concept (word level)	Advanced	Playful, not multifunction, good buy
Product Design Attribute	**Design Value**	**Design Value**
Color	Brown	Blue
Color balance	Unclear	Both Sides The Same
Color harmony	Two Colors	Two Colors
Top color	Separate	Separate
Clear	Yes	No
Grip shape	Straight	Straight
Grip quality	Rubber	Rough
Grip texture	No	Yes
Kinds of grip	Rubber	No
Mechanics	Knock	Knock
No. of mechanics	Three Colors	Three Colors
Automatic pencil	Yes	Yes
Eraser	No	No
Top material	Plastics	Metal
Top joint	Separate	Not Separate
Length	150 mm+	150 mm–
Hand grip	13 mm+	11 mm–
Weight	16 g+	16 g+
Thickness	Bold	Middle Bold
Kind of clip	Normal	Normal
Material	Plastics	Metal
Middle ring	Yes	No

by the rule-based system for the concepts *advanced*, and *playful*, and *good buy*. Two-design value sets derived from different paths in multilevel framework indicate rather different product designs. The concept *advanced* seems to propose large, functional, and plastic (mechanics-oriented) ball pens, whereas *playful* and *good buy* seems to propose smart, fashionable, and metal (Kansei-oriented) ball pens. From the result, design experts could generate some possibilities for new ball pens. Finally, we can say that the rough set model will provide a new, powerful tool in KAE, and it will have many possibilities in Kansei product design.

References

Hirata, R. O., Nishino, T., and Nagamachi, M. (2007). Comparison between statistical and lower/upper approximations rough sets models for beer can design and prototype evaluation, *Proceedings of 2nd European Conference on Affective Design and Kansei Engineering*, CD-ROM.

Kanda, N. (Ed.). (2000). *Seven tools for product planning*, Nikkagiren (JUSE).

Mori, N. (1999). Chain structure charactering Kansei engineering, *Kansei Engineering Symposium*, 12–17.

Mori, N., Tanaka, H., and Inoue, K. (Eds.) (2004). *Rough set and Kansei*, Tokyo: Kaibundo Publishing.

Nagamachi, M. (1989). *Kansei engineering*, Tokyo: Kaibundo Publishing.

Nagamachi, M. (1995). *Introduction to Kansei engineering*, Tokyo: Japan Standard Association.

Nagamachi, M. (Ed.) (2005). *Product development and Kansei*, Tokyo: Kaibundo Publishing.

Nagamachi, M. (2006). Kansei engineering and rough sets model, *RSCTC*, LNAI, 4259, Springer, 27–37.

Nishino, T. (2005). Rough sets and Kansei, in M. Nagamachi (Ed.), *Product Development and Kansei*, Tokyo: Kaibundo Publishing.

Nishino, T. (2009a). Rough set model as Kansei engineering technology and its applications, *Kansei Engineering*, 8 (1), 24–30.

Nishino, T. (2009b). Rough set model and its application to Kansei product design, *Proceedings of 10th World Congress on Ergonomics (IEA2009)*. CD-ROM.

Nishino, T., and Nagamachi, M. (2007a). Rough set model and its application to Kansei engineering, *Proceedings of 2nd European Conference on Affective Design and Kansei Engineering*, CD-ROM.

Nishino, T., and Nagamachi, M. (2007b). Kansei product development based on the extraction of multilevel decision rules to actualize customers wants using probabilistic rough set model, *Proceedings of 1st International Conference on Kansei Engineering and Emotion Research (KEER2007)*, CD-ROM.

Nishino, T., and Nagamachi, M. (2009). Kansei analysis and design of car floor mat based on rough set model. *Proceedings of 10th World Congress on Ergonomics (IEA2009)*, CD-ROM.

Nishino, T., Nagamachi, M., and Ishihara, S. (2001). Rough set analysis on Kansei evaluation of color, *Proceedings of the International Conference on Affective Human Factors Design*, Asian Academic Press, 109–115.

Nishino, T., Nagamachi, M., and Sakawa, M. (2006). Acquisition of Kansei decision rules of coffee flavor using rough set method, *Kansei Engineering International*, **5** (4), 41–50.

Nishino, T., Nagamachi, M., and Tanaka, H. (2005). Variable precision Bayesian rough sets model and its application to human evaluation data, *RSFDGrC*, LNAI 3641, Springer.

Nishino, T., Nagamachi, M., and Tanaka, H. (2006). Variable precision Bayesian rough sets model and its application to Kansei engineering, *Transactions on Rough Sets V*, 190–206.

Nishino, T., Nagamachi, M., Sakawa, M., Kato, K., and Tanaka, H. (2006). A comparative study on approximations of decision class and rule acquisition by rough sets model, *Kansei Engineering International*, Vol. **5** (4), 51–60.

Pawlak, Z. (1984). Rough classification, *International Journal of Man-Machine Studies*, 20, 469–483.

Pawlak, Z. (1998). Rough sets elements, In L. Polkowski and A. Skowron, (Eds.), *Rough sets in knowledge discovery 1*, Heidelberg: Springer.

Polkowski, L., Tsumoto, S., and Lin, T. Y. (Eds.) (2000). *Rough set methods and applications*, Springer.

Slezak, D. (2004). The rough Bayesian model for distributed decision systems, in RSCTC (Ed.), LNAI306, Springer, 684–393.

Slezak, D. (2005). Rough sets and bayes factors, *Transaction on Rough Set III*, LNCS 3400, 202–229.

Ziarko, W. (1993). Variable rough sets model, *Journal of Computer and System Sciences*, 46, 39–59.

8

Kansei/Affective Engineering and Web Design

Anitawati Mohd Lokman

CONTENTS

8.1 Introduction to Web Design

A Web site is a collection of related Web pages accessible using a domain name or IP (Internet protocol) address via the Internet or a private local area network. A Web page consists of contents such as text, images, videos, or other kinds of digital assets. Designing a Web site requires skill in creating presentations of content that is delivered ubiquitously through the World Wide Web. Typically, the basic components of Web design include content, usability, appearance, and visibility. Web site design, or Web design, may involve multiple disciplines in information systems, information technology and communication design, graphic design, human–computer interaction, information architecture, interaction design, marketing, photography, search engine optimization, and typography. Thus, to successfully produce a good Web site, a proper approach before one can start creating Web sites is required.

Web site architecture (WA) is an organized approach in planning and designing a Web site. Like the traditional architecture, WA involves technical, aesthetic, and functional criteria. Similarly, the main focus is on users and user requirements. Technical aspects of WA address the back-end components that deal with the underlying technology such as source code, data warehouse, and server-side component. It also addresses the business plan and the information architecture (IA), which deals with the structural design, method of organizing and labeling Web pages, and ways of demonstrating the Web site to the digital landscape. The aesthetic aspect addresses the user friendliness of the Web site and the visual and audible impressions. Functional aspects deal with the front-end components, which include usability, visibility/searchability, and accessibility.

8.1.1 Positioning Kansei/Affective Engineering in Web Design

In the process of designing Web sites, the aesthetic and functional criteria require particular attention to *user experience* (UX), the term used to describe the experience that a user has as a result of his or her interactions with the Web site. The consequences of UX in Web design result from the user's perception toward the elements of functional and nonfunctional qualities. Perception of functional qualities particularly lies in the usability and usefulness of the Web site. On the other hand, perception of nonfunctional qualities particularly lies in the aesthetic aspect, which includes visual and audible quality of the Web site. The most important component that cannot be neglected is the affective qualities that address the aspect of feelings, reaction, appraisal, and behavior from users. Affective qualities encompass both the components of functional and nonfunctional qualities.

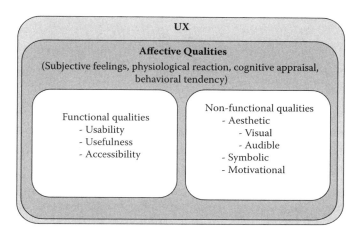

FIGURE. 8.1
Elements of UX in Web design.

In the process of designing Web sites, designers have been putting effort into what the user wants in order to come out with the desired result. However, by its very nature and similar to any kind of software system, Web design always produces conflicts regarding designer specification and user conformance. In realizing the concept in terms of Web site appearance, designers often misunderstand the description provided by users or user advocates, who in turn do not really understand the concept that they actually want. Designers possibly come out with the specification by their own intuition and creativity. Another problem is that even though the users may have well described the concept that they want, the designer's side has no clue on the design requirements to produce the kind of Web site that the users have described.

Kansei/affective engineering (KAE) provides a systematic way of understanding the insights of user perceptions toward artifacts via several physiological and psychological measurement methods. These insights are then translated to the design characteristic of the artifact. This approach of KAE matches the concern of requirement specifications to correspond to the foundation of UX in Web design, providing the possibility for users to express their concept and providing clues to designers in the form of design requirements so that the designer can objectively develop the desired Web design. In every aspect that contributes to the desired UX, where the design needs to match the insights of the users' experience, KAE is seen as a possible requirement generation technique. The implementation of KAE in every possible aspect of UX in Web design will enable the identification of the concept of Kansei in Web design and the contributing design specification to the concept in the form of design requirements.

8.2 Engineering Kansei in Web Design

This chapter attempts to demonstrate the implementation of KAE in Web design, focusing on the visual quality of design that shapes one of the elements that have consequences on the user experience. The detail of the implementation is described in the following sections, beginning with the building of a reference model for the implementation of KAE in Web design.

8.2.1 Structuring the KAE Method into Kansei Design Model

From the review of KAE literature, a gap in terms of the description of steps to be performed in the implementation method of Kansei engineering was identified. There are many types of techniques in different kinds of implementation, but the description of the method is largely narrative. For this reason, based on previous literature involving the adoption of KAE, setting the foundation to the basic principles of KAE, the method is structured into a model called the *Kansei design model*. In structuring the model, careful attention was given to the capacity and availability of infrastructure, facilities, and cost. The structured model employs a self-reporting method in the measurement of Kansei, allowing KAE implementation in a basic environment where no special equipment and skills are required. With this model, the audience can have a useful guide to the implementation of KAE.

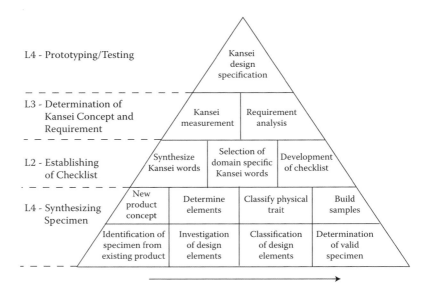

FIGURE 8.2
The Kansei design model.

The model is developed to provide a systematic approach to the implementation of KAE in designing Kansei products. The model presented is a useful mechanism for industries, designers, academic researchers, and other stakeholders in discovering Kansei concept, and design requirements for the development of Kansei products. The model is divided into four levels: L1, L2, L3, and L4. The following sections describe details of each level.

8.2.1.1 L1: Synthesizing the Specimen

L1 is the level of synthesizing the specimen. The level is subdivided into two different procedures, PI and PII. These procedures differ in the process of synthesizing the specimen. There are four steps in both PI and PII, which are essential in determining a valid specimen. The procedure can be decided according to one's objective.

PI is applicable to products that already exist in the market where the maker needs to improvise the design. The procedure begins with a collection of samples with visible differences from existing products in the market within a specific domain. KAE emphasizes controlling the domain, as the consumer's response is unique with different domains (Nagamachi, 2003; Ishihara et al., 2005). Previous KAE studies have suggested different techniques for determining specimens from using actual products and using pictures of product. The choice of specimen depends on its suitability to the experimental design.

Then, the following procedure is the process of investigating design elements in all samples. Determination of the number of design elements depends on the level of detail that needs to be included in one study. Controlling the number of elements enables a more objective measurement. On the other hand, including all identifiable elements from the consumer's point of view enables a more accurate measurement. The latter ensures the accuracy of design requirements as an outcome of a study, as consumers are assessing a product as a whole. To match the consumer's emotional response to design elements, this chapter suggests that controlling the number of elements will produce a less accurate result.

The next procedure is the classification of design elements. The identified design elements are further classified into item and category. *Item* is the type of common physical traits of all specimens such as background color, body shape, and text alignment. *Category* is the specific attribute of the item in each specimen, such as red as a background color of specimen A, and blue as a background color of specimen B. The process is crucial since the findings will be the essence in the success of the requirement analysis stage. Finally, based on a set of rules, a valid specimen for Kansei measurement can be synthesized among all the initial samples.

PII, on the contrary, is designed for application when a company or designer plans to design a new concept of a product based on their objectives. This is applicable to the development of product that has yet existed on the market. In this case, designers and experts have to determine product specifications based on their inspiration in relation to the target concept. For instance, to design an *elegant* mobile phone, the process begins with synthesizing words related to the concept of *elegant* within the domain. Then, designers or experts have to determine design elements that have a connection with an elegant feeling, classify the physical traits, and build a number of prototypes based on the technical specification. This prototype will then be used as a specimen at the following level of the model to confirm their design with consumers.

8.2.1.2 L2: Establishment of Kansei Checklist

L2 describes preparation and establishment of Kansei checklist. The level is divided into three steps: (1) synthesizing Kansei words, (2) selection of domain-specific Kansei words, and (3) development of checklist. The level synthesizes Kansei words, from a larger number of possible Kansei words to focused Kansei words that highly related to the product domain. Kansei words can be adjectives, such as *calm*, *sophisticated*, and *natural*, or nouns. These Kansei words can be synthesized from pertinent literature, technical magazines, or even consulting experts. Finally, utilizing the Kansei words, L2 produces a Kansei checklist in the form of the semantic differentials (SD) scale as a measurement tool for Kansei measurement in the next level.

8.2.1.3 L3: Determination of Kansei Concept and Requirement

L3 describes determination of the Kansei concept and requirement. This level is divided into two steps: (1) Kansei measurement, and (2) requirement analysis. In the first step, Kansei measurement is performed using expert or ordinary consumers as test subjects. The subjects are required to rate their impressions toward product specimen on the Kansei checklist. Results from the evaluation will be analyzed to interpret links between subjects' Kansei and design elements identified in L1. The outcome can be used to determine design requirements for the development of the Kansei product.

8.2.1.4 L4: Prototyping/Testing

L4 describes prototyping/testing. In this final level, the results from L3 will be used as the foundation to build a prototype of the Kansei product. The process will involve the employment of the Kansei concept and design requirements identified in L3. To develop a successful Kansei product, experts' creativity should be included in the design process. Testing may be performed to validate the design requirements.

8.3 Demonstration of the Kansei Design Model Implementation in Web Design

This section demonstrates the implementation of the constructed Kansei design model in Web design to explore the potential of Kansei engineering in Web design.

8.3.1 Synthesizing the Specimen

In order to implement KAE in Web site interface design, the first procedure, PI, of the Kansei design model is adopted. The procedure is applicable in engineering the Web site interface designs and provide guidelines for designers to improvise Web site interface design into Kansei design. This section focuses on the context of interface design that is visible to users of clothing Web sites.

Four stages are involved in implementing level 1 of the model:

1. Identification of initial specimens
2. Investigation of design elements
3. Classification of item and category
4. Finalizing valid specimen

8.3.1.1 Identification of Initial Specimens

Table 8.1 shows the controls and criteria used to select initial specimens as part of preparation of the instruments to be used in the Kansei measurement process.

In identifying the initial specimens, 163 Web sites were selected based on their visible design differences in both content and layout context (color, typography, layout, etc.). These Web sites were chosen according to their listing on the Apparel Search Web site (http://www.apparelsearch.com). Apparel Search is the leading online clothing directory and has the categorization structure that helped in selecting Web sites.

8.3.1.2 Investigation of Design Elements

Each component within the basic structure of the Web site was used as the basis during the empirical investigation of all design elements that compose all of the 163 specimens. The design elements are broken down into item and category so that clear categorization of designs can be organized. In total 77 items have been identified from the empirical investigation. These items

TABLE 8.1

Control Condition

No.	Item	Condition
1.	Web site criterion	Visible differences in design
2.	Focus context	Design content and layout
3.	Screen resolution	1024 × 768 pixels
4.	Access/download date	June 1–30, 2006
5.	Platform	Win32
6.	Operating system	Windows XP
7.	Color quality	32 bit
8.	Browser	Opera 9.00
9.	Browser control	Encoding = Windows-1252
		Default language = English, [en]
		Default Text Size = Medium
		Colors = Windows 32 bit color
10.	Encoding	Windows-1252
11.	Language	English
12.	Default text size	Medium (3 pt)

were then investigated individually from each Web site to identify categories assigned to each. Table 8.2 summarizes all design elements in the 163 Web sites that are transparent from the viewpoint of Web site visitors.

8.3.1.3 Classification of Item and Category

In the context of the basic structure of a Web page, each specimen may comprise all or part of the elements within each section: Body (refers to the layer that a Web page may reside on), Page (refers to the page of a Web site), Header (refers to the head section of the Web page, which may contain the menu), Top Menu (refers to the menu on the top of the page), Left Menu (refers to the menu on the left pane of the Web page), Main (refers to the main body of the Web page), Right Menu (refers to the right menu on the right pane of the Web page), and Footer (refers to the footer of the Web page, which may contain a menu). The Web page may also contain pictures and other elements such as artistic menus and logo. From the set of items identified in the previous section, the classifications of categories are identified that form the different characteristics of Web site designs. Table 8.3 gives examples of item and category to be investigated from all specimens. A total of 77 item and 249 category of Web site design elements were identified.

To simplify the organization of the huge amount of data, all the identified design elements (item and category) were organized into specimens by design elements matrix. Each specimen was carefully investigated to check the item and category that make up the characteristic of the specimen.

TABLE 8.2

Item Identified from the Initial Specimens

Section	Item
Body	Background color, background style
Page	Shape, menu shape, style, orientation, color, size, border existence
Header	Existence, background color, background picture existence, font size, menu existence, menu link style, menu background color, menu font size, menu font family, menu font style
Main	Background color, background picture existence, shape, adv. Existence, text existence, text alignment, font color, font size, font family, font style
Top menu	Existence, location, link style, background color, font color, font size, font family, font style
Right menu	Existence, style, font size
Left menu	Existence, link style, background color, font color, font size, font family, font style
Footer	Existence, menu existence, menu link style, menu background color, menu font color, menu font size, menu font family, menu font style, shape
Picture	Existence, size, dimension, focus, arrangement, style, image used?, No of people in 1 picture, body representation type, face expression, face facing? Empty space? Other images? Product display style, product try on? Product view style
Others	Dominant item, artistic menu used? Discount advertisement existence, logo existence, logo location

TABLE 8.3

Example of Design Item and Category

Item	Category
Page background color	Blue
Left menu style	Button
Main text size	Medium

When the category matched the investigated specimen, then the matrix was checked. The process is repeated until investigation of the 249 categories in all specimens is completed. Although the construction of a matrix does not substantially reduce the amount of work, which is impossible, it offers easy management of the knowledge by providing orderly data organization. The matrix data also alleviate the screening procedure, involving 249 categories in 163 specimens, for identifying a valid specimen.

8.3.1.4 Finalizing Valid Specimen

Results from the item/category classification stage were then examined according to the following rules in conformance to KAE methodology

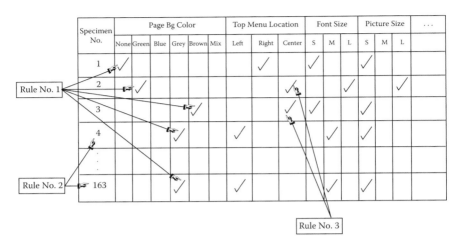

FIGURE 8.3
Rules in screening specimens.

1. For each sample, only one category under one Item is ticked.
2. Only one sample will be taken if exactly the same category under each Item is ticked for more than one Web site.
3. Take two or more sample Web sites where same category is ticked.

Figure 8.3 illustrates how the rules are executed in screening the Web site specimen to identify valid samples.

Conforming the first rule, for Page Bg Color every specimen must only have one color checked. Secondly, specimens 4 and 163 have exactly the same result, so only one can be included as a valid sample. Finally, two or more samples of the same category, for example Top Menu Location and Center in specimens 2 and 3, must be included. This simple set of rules has helped narrow down the previously identified 163 initial specimens into a smaller number to be used as valid specimens in the empirical studies. Although the rules followed are simple, the screening of 249 categories of more than 163 specimens was enormously demanding. With careful attention, 35 Web site specimens were finally concluded. The specimens were coded numerically from 1 to 35. A snapshot of the specimens is shown in Figure 8.4.

8.3.2 Establishment of Checklist

This section develops the Kansei checklist as one of the instruments to be used in Kansei measurement and describes the stages of the establishment of the checklist.

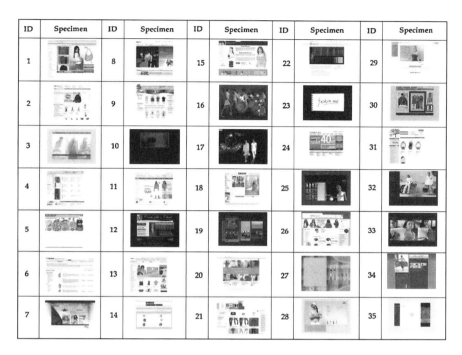

ID	Specimen	ID	Specimen	ID	Specimen	ID	Specimen	ID	Specimen
1		8		15		22		29	
2		9		16		23		30	
3		10		17		24		31	
4		11		18		25		32	
5		12		19		26		33	
6		13		20		27		34	
7		14		21		28		35	

FIGURE 8.4
The 35 valid specimens.

8.3.2.1 Checklist Development

A checklist is developed as one of the instruments used as measurement tools. The checklist comprises Kansei words that are identified according to the steps described in the following sections. The Kansei words are used as the measure of strength of the Kansei, that is, the emotional responses that subjects feel when looking at the Web site.

8.3.2.2 Synthesize Kansei Words

A set of Kansei words was selected based on frequency of appearance in Web design guidebooks, Web sites, research papers, and journals. Additionally, general Kansei words were added according to relevance in describing Web sites. A total of 40 words were then finalized to be used in the experimental procedure. Some of the synthesized words were *adorable, classic, creative, elegant, lovely, masculine,* and *sophisticated.*

8.3.2.3 Development of Kansei Checklist

The 40 Kansei words compiled in the earlier section were then organized into a 5-point SD scale to form the Kansei checklist. This checklist will be used as a measurement tool for investigating users' Kansei.

8.3.3 Determination of Kansei Concept and Requirement

In determining the Kansei concept and design requirement in Web design, two stages of activities were performed. The following subsections describe the stages.

8.3.3.1 Kansei Measurement

The first stage describes the process of Kansei measurement that is performed to evaluate visitors' Kansei responses when interacting with Web sites. This stage consists of the core activities involved in engineering emotion in Web site design. The instruments and equipments used in the Kansei measurement procedure are the Kansei checklist, 120 test subjects, the screenshot of the 35 valid specimens, a computer, and one large LCD screen.

1. Instrumentation. The instruments involved in the experimental procedure are as follows:
    a. The 35 valid specimens. The specimens were coded numerically from 1 to 35. A snapshot of all specimens can be found in Figure 8.4.
    b. The Kansei checklist, which consists of 40 Kansei words organized in a 5-point SD scale. The order of Kansei words was organized into five different arrangements to minimize bias in the evaluation process. A full Kansei checklist can be found in Appendix 5.7.
    c. One hundred and twenty test subjects. Subjects for the empirical study were employed with equal distribution of male and female subjects of 30 people in four groups. A total of 120 undergraduate students from

       - Information Technology and Quantitative Science Faculty (IT)
       - Architecture, Building, Planning, and Survey Faculty (AD)
       - Business and Management Faculty (BM)
       - Engineering Faculty (ER)

       from the author's university were recruited for the empirical evaluation. All of them were in their 20s, Internet users, and familiar with online shopping. The suitability of the employment of young students as subjects in this research is supported by the literature in information system studies that suggests students and youngsters represent the majority of e-commerce consumers (Saarenpää and Tiainen, 2005). Therefore, they are the best consumer demographic group to be studied. On the other hand, the population of subjects in this research was decided based on the suggested number in KAE methodology. Although the population of test subjects varies from a minimal number such as five to more than a thousand in different KAE implementation, depending on objectives and measurement tools used, the suggested number for this kind of consumer research is around 30–50 subjects (Nagamachi, 2003; Nagamachi and Lokman, 2010).

In the research, in an effort to explore differences in educational background, a total of 120 students from four different academic backgrounds were chosen. Additionally, to facilitate balance of gender population, equal distribution of subject numbers—15 females and 15 males—was employed in each group.

2. Evaluation. Four Kansei evaluation sessions were held separately for each group. During each session a briefing was given before the subjects began their evaluation exercise. The 35 Web site specimens were shown one by one on a large white screen to all subjects in a systematic and controlled manner. Subjects were asked to rate their feelings on the checklist according to the given scale. Subjects were given 3 minutes to rate their feelings toward each specimen. They were given a break after the 15th Web site specimen, to refresh their minds. The order of the checklist was also changed to eliminate bias. Each Kansei evaluation session took approximately 2 hours to complete.

8.3.3.2 Requirement Analysis

This stage is performed to conceptualize Kansei in Web site design and to analyze design requirements. Multivariate analyses were performed to find empirical evidence toward the goal of engineering emotion in Web design. The outcomes of the analyses were used to propose a guideline to the design of a Kansei Web site. We calculated the average Kansei evaluation value of each sample obtained from all subjects from the experimental procedure. These averaged data from the evaluation results were used in the calculation of the multivariate analyses.

Factor analysis (FA) was used to analyze Kansei concept, and partial least squares analysis (PLS) was used to analyze relationships of design and Kansei.

8.3.3.2.1 The Kansei Concept

The research performed FA to determine the concept of Kansei in Web design. Table 8.4 shows the result of FA after varimax rotation. Varimax rotation, which was originated by Kaiser (1958), is the most popular

TABLE 8.4

Factor Contribution

Factors	Variance	Contribution	Cumulative Contribution
Factor 1	16.09262	40.23%	40.23%
Factor 2	12.29421	30.74%	70.97%
Factor 3	3.427578	8.57%	79.54%
Factor 4	1.856272	4.64%	84.18%
Factor 5	1.810882	4.53%	88.70%
Factor 6	0.923415	2.31%	91.01%
Factor 7	0.370649	0.93%	91.94%
Factor 8	0.250962	0.63%	92.57%

rotation method that simplifies the interpretation of variables. In the table, it is evident that the first factor explains 40.23% of the data and the second factor explains 30.74% of the data. Both factors represent the majority of factor contributions. This shows that Factor 1 and Factor 2 have a dominant influence on Kansei words. The first two factors together represent 70.97% of the variability, while three factors explain 79.54% of the variability. Thus, the research considered including the third factor to increase the proportion that represents most of the data. The proportion of variability explained by the fourth factor and the rest are minimal (4.64% and less), and they probably can be ignored as they can be considered insignificant.

Table 8.5 shows factor loading results after varimax rotation. The table shows factor results in ascending order. The structure of Kansei words is observable from the table. Variables that have a high score are perceived as significant factors in Web site design. The research set approximately 0.7 as the reference score and cross-checked with the result of correlation coefficient analysis (CCA) to draw conclusions. The fourth and fifth factors were included since they can be regarded as having a clear image on the Web site Kansei, even though the score is slightly lower. Henceforth, the conclusion from the results of FA was that the Kansei concepts of Web design are structured by five factors: *sophistication, elegant-beauty, simplicity, lightness,* and *tidiness*. These five factors altogether explain 88.70% of the total data.

The first factor, *sophistication*, consists of mystic, futuristic, masculine, luxury, sophisticated, surreal, impressive, gorgeous, cool, and professional. The second factor, *elegant-beauty*, consists of feminine, chic, beautiful, cute, sexy, charming, adorable, and elegant. The third factor, *simplicity*, consists of simple and plain. The fourth factor, *lightness*, consists of light. The fifth factor, *tidiness*, consists of neat and natural.

As can be seen from the result, the first and second factors explain most of the data; *sophistication* and *elegant-beauty* represent 70.97% of data. Thus, these two factors are very important Kansei concepts. This indicates that all

TABLE 8.5

Factor Loading for Kansei Words

Variable	Factor 1	Variable	Factor 2	Variable	Factor 3	Variable	Factor 4	Variable	Factor 5	Variable	Factor 6
Professional	0.805803	Pretty	0.689458	Sexy	0.272922	Impressive	0.202914	Lively	0.201622	Adorable	0.046115
Cool	0.811333	Lovely	0.690027	Classic	0.275441	Masculine	0.211368	Lovely	0.207907	Creative	0.055077
Gorgeous	0.812754	Elegant	0.703414	Boring	0.308598	Adorable	0.21181	Beautiful	0.239447	Classic	0.063075
Impressive	0.822734	Adorable	0.713039	Light	0.313839	Cool	0.284239	Light	0.247409	Light	0.066423
Surreal	0.846445	Charming	0.763686	Neat	0.319281	Interesting	0.308149	Relaxing	0.274663	Plain	0.066736
Sophisticated	0.848426	Sexy	0.787619	Calm	0.339163	Comfortable	0.320447	Calm	0.302602	Chic	0.131081
Luxury	0.878831	Cute	0.794058	Relaxing	0.348516	Lively	0.328505	Comfortable	0.328556	Old-fashioned	0.144072
Masculine	0.899118	Beautiful	0.816958	Natural	0.424887	Refreshing	0.390941	Refreshing	0.354777	Fun	0.222652
Futuristic	0.913165	Chic	0.93916	Plain	0.839005	Natural	0.49998	Natural	0.604973	Cute	0.276694
Mystic	0.941857	Feminine	0.948707	Simple	0.9241	Light	0.610599	Neat	0.738318	Childish	0.633318

Web sites should have these two factors in order to produce optimum results. *Simplicity*, *lightness*, and *tidiness* are also important but have weaker influence. Therefore, it is suggested that these factors be used as background/supporting elements in Kansei Web site design.

8.3.3.2.2 Analyzing the Relationship of Design and Kansei

In analyzing design requirements, PLS analysis is used to link the Kansei responses with design elements. In this analysis, there are three sets of data used:

1. The dependant (objective) variables, y, that is, the 40 sets of Kansei responses by 120 subjects.
2. The sample, s, that is, 35 Web sites.
3. The independent (explanatory) variables, x, the design elements (design category).

As mentioned earlier, the investigation of design elements has resulted in 77 design items composed of 249 categories. For PLS analysis purposes, all these categorical variations were converted into dummy variables. The result of averaged data, for example, *adorable* data, was then appended to the next column right after the last column of design category. This research has 40 predictors and therefore the analysis was repeated 40 times, exchanging the predictor into the last column. An instance from the result for the Kansei score by design category can be found in Table 8.6.

The research analyzed the result of the PLS coefficient score to determine relations between Kansei and design elements. In order to determine the influence of design elements to Kansei, PLS *Range* for each Kansei was calculated. The calculation of *Range* enables the identification of design influence, good design and bad design. *Range* is calculated using maximum and minimum value, where

$$Range = \mathrm{PLS}_{\mathrm{Max}} - \left|\mathrm{PLS}_{\mathrm{Min}}\right|$$

Mean of *Range* is calculated, where

$$\overline{Range} = \frac{1}{n}\sum_{i=1}^{n} Range_i$$

Each Kansei has means of *Range*, and if the mean value of a category is larger than $\overline{Range}$, the item is considered to have good influence in design. *Range* for every category having value larger than $\overline{Range}$ implies a best fit category that highly influences users' Kansei in Web site design.

TABLE 8.6

PLS Coefficient Values

Category	Kansei						
	Adorable	Appealing	Beautiful	Boring	Calm	Charming
BodyBgColor-White	−0.0365510	−0.03699	−0.01674	0.024457	−0.02534	−0.0355
BodyBgColor-Black	0.0065448	0.011992	−0.01374	−0.00265	0.028478	0.005989
BodyBgColor-DKBrown	0.0604354	0.067045	0.018645	−0.03459	0.034535	0.062087
BodyBgColor-LtBrown	0.0132480	0.011571	−0.00476	0.006006	0.017753	0.021147
BodyBgColor-Gray	0.0293157	0.036547	0.050832	−0.044	0.006308	0.033964
BodyBgColor-LtBlue	0.0214068	0.004199	0.01559	−0.01155	−0.01207	0.005272
PageMenuShape-Curve	0.0005942	−0.01064	0.006802	−0.00788	−0.00697	−0.01168
PageMenuShape-Sharp	−0.0122160	−0.00199	−0.02235	0.021834	0.006232	−0.00213
PageMenuShape-Mix	0.0286988	0.024137	0.042154	−0.03895	−0.00256	0.026368
PageStyle-Frame	0.0340362	0.025436	0.027154	−0.03955	0.005176	0.005524
PageStyle-Table	−0.0420300	−0.03508	−0.02236	0.04195	−0.01811	−0.01117
PageOrientation-BC	−0.0447260	−0.04155	−0.03098	0.026269	−0.03299	−0.0298
PageOrientation-Content	0.0326846	0.037362	0.004381	0.011088	0.024659	0.034727
PageOrientation-Header	−0.0546260	−0.05817	−0.03232	0.037364	−0.03451	−0.06225
PageOrientation-HF	0.0157878	0.015311	0.026275	−0.01806	0.00773	0.014688
PageOrientation-H Split	0.0228924	0.02168	0.004236	−0.01846	0.016584	0.022154
PageOrientation-V Split	0.0189942	0.034075	0.030549	−0.02336	0.016852	0.028757
PageOrientation-Plain	0.0241808	0.019132	0.007982	−0.01443	0.015297	0.005532
DominantItem-Pict	0.0467300	0.048044	0.030602	−0.04358	0.014746	0.050762
DominantItem-Adv.	−0.0296800	−0.03225	−0.01741	0.019399	−0.00744	−0.02078
DominantItem-Text	−0.0561230	−0.04549	−0.02663	0.050166	−0.00284	−0.04293

TABLE 8.7

Design Influence in Kansei

| Design | Adorable | | Appealing | |
Influence No.	Design Element	Range	Design Element	Range
1	Page Color	0.11488	Header Bg Color	0.12338
2	Product Display Style	0.10644	Face Expression	0.12216
3	Header Menu Bg Color	0.10612	Header Menu Bg Color	0.12077
4	Left Menu Font Color	0.10370	Product Display Style	0.10646
5	Header Bg Color	0.10218	Body Bg Color	0.10574
6	Face Expression	0.10024	Page Color	0.10091
7	Body Bg Color	0.10015	Left Menu Font Color	0.10085
8	Dominant Item	0.09980	Picture Style	0.09771
9	Header Font Size	0.09651	Page Orientation	0.09182
10	Main Text Existence	0.08813	Dominant Item	0.09141
11	Main Bg Color	0.08587	Main Text Existence	0.08811

In every category, maximum value shows best fit value of design elements that influence the Kansei, and minimum value shows the worst value of design elements that influence the Kansei.

Table 8.7 shows an example of the result of the selected category for which *Range* has value larger than $\overline{Range}$, for each Kansei. The results are sorted in descending order to illustrate dominant design category for each Kansei.

The result shows that in designing an *adorable* Web site, the designer must set priorities to design elements according to higher order of influence, such as page color, product display style, header menu background color, left menu font color, and so forth. On the other hand, in designing an *appealing* Web site, the designer must set priorities to design elements according to higher order of influence, such as header background color, face expression, header menu background color, and so forth.

8.3.3.2.3 Proposing Kansei Web Design Guidelines

All the above analyses have enabled the research to propose Kansei Web Design Guideline, a guideline to the design of Kansei Web sites. Results of Kansei structure from FA are used to conceptualize Kansei, and results from PLS scores are used to compose the design requirement. The design requirements included in the guidelines are the elements that have high influence in eliciting each Kansei. Table 8.8 shows examples of the results of the established guideline.

To effectively utilize the guideline, the audience and especially designers are advised to select the best combination possible from a Kansei concept that may consist of one or more Kansei elements. It should be noted that it is important to blend designers' creativity with the guideline to ensure the

TABLE 8.8

Customer Needs Direction Effect Measure (E)

Customer Needs Direction	Supp	Cer	Eff
Simple and Advanced and Beauty	41	0.569	1.387
Simple and Advanced and Young	41	0.569	1.387
Advanced and Original	77	0.535	1.303
Handiness and Original and Young and Functional	50	0.521	1.269

success of the Kansei product (Nagamachi, 2008), in this case an e-commerce Web site. To illustrate an example of the guideline, design elements for the Kansei *mystic* should be interpreted as follows:

Body background color should be in black.
Body background style should be in texture of color tone.
Page shape should be not specific.
Page orientation should be plain.
Dominant item should be picture, and so on.

In KAE, the success of a Kansei design product relies on the idea implied from the result of Kansei evaluation, blended with technical expertise of the designer. In the design process, the guideline should be referenced by and technical specifications must be provided by the expert.

8.3.4 Prototyping/Testing

The chapter performed two stages of activities for the purpose of validation of the successful implementation of the Kansei Design Model in engineering Kansei in Web design. The chapter developed several prototypes using the proposed guideline, and then conducted Kansei evaluation to see if there are any difference in Kansei responses after the implementation of the guideline.

8.3.4.1 Prototyping

In the KAE approach, a designer will design a new product based on a Kansei concept identified from FA results. In the case of this research, in performing confirmatory study, the research attempts to design a Kansei Web site to be used as a specimen. Although a combination of Kansei factors are ideal, in the case of Web site design, due to the large number of design variables, it is almost impossible to combine Kansei factors. For confirmatory purposes, such attempt was conducted where five Kansei were selected and the individual guideline then was used as a foundation for design specimens. The five Kansei were selected from factors 1, 2, and 3 and referred to the pro-

TABLE 8.9

The Selected Kansei Guideline

Design Element	Kansei				
	Cute	Feminine	Luxury	Masculine	Simple
Body Bg Color	Light Blue	Light Blue	Black	Black	Dark Brown
Body Bg Style	Texture	Texture	Color Tone	Color Tone	Picture
Page Shape	Sharp	Sharp	Sharp	N/S	Sharp
Page Menu Shape	Mix	Mix	Sharp	Sharp	Sharp
Page Style	None	None	None	None	None
Page Orientation	Footer	Footer	Vertical Split	Header	Content
Dominant Item	Picture	Picture	Picture	Picture	N/S
Page Color	Gray	Pink	Black	Blue	Brown
Page Size	Small	Small	Small	Medium	Medium
Other Images?	Animal	Animal	Animal	Animal	Kids
Product Display Style	Filmstrip	Filmstrip	Filmstrip	Filmstrip	Catalog
Product Try On	Yes	Yes	Yes	Yes	Yes
Product View Angle	Rear	Rear	Mix	Side	None
Artistic Menu?	Yes	Yes	Yes	Yes	No
Empty Space?	Less	Less	Less	Less	More
Discount Ad. Existence	No	No	No	Yes	Yes

posed guideline presented earlier. Table 8.9 shows an example of the selected Kansei and guideline.

The guidelines were carefully followed in the development of prototypes to be used as specimens in the confirmatory study. Figure 8.5 shows an example of the developed *cute* Web site to give an illustration of how the guideline is used in the formation of the Web site. It should be noted that the guideline is to be used to support the designer's creativity by providing design requirements to the anticipated Kansei.

Other than the above Web site, the research has developed four prototypes, and altogether five prototypes were developed to be used as a specimen in the experimental procedure. Figure 8.6 shows a snapshot of the developed five specimens.

8.3.4.2 Testing

To test the success of the guideline implementation in the prototype, a comparative study was conducted using two sets of data:

1. The exploratory data—The data extracted from initial study described in earlier sections.
2. The confirmatory data—The data obtained from Kansei evaluation performed using the prototype.

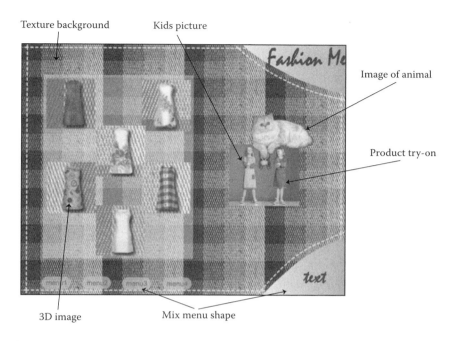

FIGURE 8.5
Example of design requirement for *cute* Web site.

ID	KANSEI	SPECIMEN	ID	KANSEI	SPECIMEN
1	CUTE		4	MASCULINE	
2	FEMININE		5	LUXURY	
3	SIMPLE				

FIGURE 8.6
Confirmatory Web site specimen.

8.3.4.2.1 Extracting the Exploratory Data

Exploratory data are the data extracted from exploratory study results. In the process of identifying which data to extract, the research performed the following procedures:

1. Data coding to segregate all data from 15 good subjects
2. Random generator to select 5 specimens
3. Calculate averaged evaluation value from the 15 good subjects

In the process of extracting data, the Microsoft Excel random generator was used to select the five specimens. Exploratory data were finalized by filtering all the data using the generated sample ID number. Average data were calculated using the filtered data and labeled as the *exploratory data* to be used in the comparison procedure.

8.3.4.2.2 Obtaining Confirmatory Data

To obtain confirmatory data, Kansei evaluation was performed with the newly developed prototype. A checklist comprising five sets of the selected Kansei was developed as a measurement tool to be used in the experiment. Using the five prototypes, the Kansei checklist, and 15 good subjects, the research performed a confirmatory Kansei measurement. The 15 good subjects were selected among good respondents from the initial study. One Kansei evaluation session was conducted to measure Kansei responses from all subjects. During the session a briefing was given before the subjects began their evaluation exercise. The five Web site specimens were shown one by one on a large white screen to all subjects in a systematic and controlled manner. Subjects were asked to rate their feelings on the checklist according to the given scale. Subjects were given 3 minutes to rate their feelings toward each specimen. The Kansei evaluation session took approximately 15 minutes to complete.

The average evaluation result obtained from the Kansei evaluation process was calculated, and the data were labeled *confirmatory data*.

8.3.4.2.3 Comparison of the Exploratory and Confirmatory Data

To provide proof that the implementation of the proposed guideline is successful in producing a Kansei Web site, the chapter performed PCA on both *exploratory data* and *confirmatory data*. Comparison of the Kansei structure formed by the two groups of data also helped confirm the validity of the Kansei design model. Both purposes can be achieved when improvement in the structure of Kansei is evident.

In performing the comparative study, the averaged data from the two groups, *exploratory data* and *confirmatory data*, were combined. PCA was then performed to investigate the structure of Kansei before and after the guideline implementation. Figure 8.7 illustrates the procedure. In the procedure,

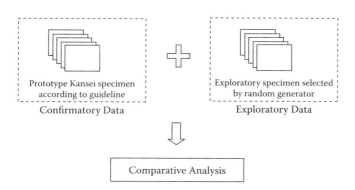

FIGURE 8.7
The testing method.

the *confirmatory data* were appended at the end of the *exploratory data* to make it easier to recognize specimens in the investigation process. In the combination, exploratory specimens were coded from 1 to 5, and confirmatory specimens were coded from 6 to 10. The overall data of the 10 samples were analyzed using PCA to investigate relations between all samples and Kansei.

Figure 8.8 shows PC vector for the *comparison data*. The vector plot shows the implied Kansei structure by all specimens in two-dimensional spaces. It is evident from the vector plot that specimens 1, 2, 3, 4, and 5, which are specimens from the exploratory study, are concentrated toward the center

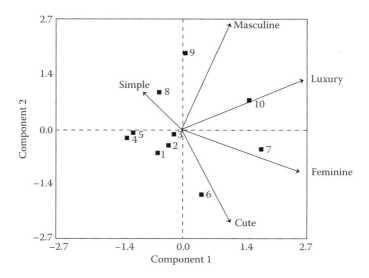

FIGURE 8.8
PC vector for comparison data.

space. This indicates that the specimens have poor influence on Kansei in comparison to the other groups. Specimens 6, 7, 8, 9, and 10, which are newly developed Kansei Web site prototypes designed based on the guideline, are very well spread all over the Kansei vector. This result provides evidence that the newly designed Web site has a good fit to Kansei. Thus, it can be concluded that the Kansei in Web site design has been improved and that the Kansei Web Design Guideline that was referenced is valid and justified.

8.4 Summary

This chapter has explored the potential of KAE implementation in Web design. Literature about product emotion has described the need to design products that capture the user's attention by captivating emotional connectivity with the interface of the product. Many products that were designed geared to the user's emotion have been produced successfully in the market (Nagamachi, 2008). With this motivating factor, the chapter explored the possibility of embedding users' emotion in Web site design. When addressing the diverse aspects of UX in Web design, KAE can be seen to provide a systematic approach to discover the concept of user Kansei in Web design, and thus enable designers to strategize the design of a Web site that caters to the insight of user emotions and feelings in its design by using KAE, the emotion.

Until recently, the design requirement for Web sites has been focused on the aspect of cognitive functionality and usability (Backlund, 2001; Garret, 2003; Ivory and Hearst, 2001; Krug, 2000; Lederer et al., 1998; Marcus and Gould, 2001; Nielsen, 2000). Although recent research has paid increasing attention to the emotional aspect of Web site design (Kim et al., 2003; Li and Zhang, 2005; Norman, 2002; Overbekee et al., 2004; Thielsch, 2005), research is lacking in terms of determining the aspects of emotional design concept and requirements. Consequently, paradigm designs still must include the emotional aspect of Web site design according to designer's interpretation and inspiration. Unfortunately this does not take into account the user's implicit needs and emotions. Thus, this chapter has attempted to engineer emotion in Web site design to fill in the gap of design requirements geared to users' emotional responses.

The chapter has provided evidence that KAE can be used to determine the implicit needs and emotion of users and shape the concept of Kansei in Web design. The chapter has also formulated design requirements to create the desired concept of Kansei Web design. Users or customers will then have a guide to objectively describe the concept of Web design that they view or want, and the designer will have clues for designing such a Web site.

References

Backlund, J. (2001). *Web Interfaces and Usability*. Stockholm: Center for User Oriented Design, ISSN 1403-0721.

Garrett, J. J. (2003). *The Elements of User Experience*. New York: New Riders.

Ishihara, I., Nishino, T., Matsubara, Y., Tsuchiya, T., Kanda, F., Inoue, K. (2005). Kansei and Product Development, in M. Nagamachi (Ed.), Vol. 1. Tokyo: Kaibundo Publishing.

Ivory, M .Y., Hearst, M. A. (2001). The State of the Art in Automating Usability Evaluation of User Interfaces, *ACM Computing Surveys* 33, 4: 470–516.

Kaiser, H. F. (1958). The Varimax Criterion for Analytic Rotation in Factor Analysis, *Psychometrika*, 23, 187–200.

Kim, J., Lee, J., Choe, D. (2003). Designing Emotionally Evocative Homepages: An Empirical Study of the Quantitative Relations between Design Factors and Emotional Dimensions. *International Journal of Human-Computer Studies*, 56, 6: 899–940.

Krug, S. (2000). *Don't Make Me Think*, New York: New Riders.

Lederer, A. L., Maupin, D. J., Sena, M. P., Zhuang, Y. (1998). The Role of Ease of Use, Usefulness and Attitude in the Prediction of World Wide Web Usage. In S. P. Robbins, *Organizational Behavior*, 8th edition. Upper Saddle River, NJ: Prentice Hall, p. 168.

Li, N., Zhang, P. (2005). Towards E-Commerce Web sites Evaluation and Use: An Affective Perspective. Post-ICIS '05 JAIS Theory Development Workshop, Las Vegas, NV.

Marcus, A., Gould, E.W. (2001). Cultural Dimensions and Global Web Design: What? So What? Now What? White Paper, AM+A.

Nagamachi, M. (2003). *The Story of Kansei Engineering*, Tokyo: Japanese Standards Association.

Nagamachi, M. (2008). A successful statistical procedure on Kansei engineering product, The 2nd European Conference on Affective Design and Kansei Engineering, (Helsingborg). (CDROM).

Nagamachi, M., Lokman, A. M. (2010). *Innovations in Kansei/affective engineering*. Industrial Innovation Series, A. B. Badiru (Ed.). (in press), Boca Raton, FL: CRC Press.

Nielsen, J. (2000). *Designing Web Usability: The Practice of Simplicity*. New York: New Riders Press.

Norman, D. A. (2002). Emotional Design: Attractive Things Work Better. *Interactions: New Visions of Human-Computer Interaction*, ix, 36–42.

Overbeeke, K., Djajadiningrat, T., Hummels, C., Wensveen, S. (2004). Beauty in Usability: Forget About Ease of Use! Human-Computer Interaction.

Saarenpää, T., Tiainen, T. (2005). *Empirical Samples of IS Studies on eCommerce Consumers*, IRIS28. Norway.

Thielsch, M. T. (2005). *Web-evaluation: Aesthetic Perception of Web sites*, University of Münster, Germany.

9

Kansei/Affective Engineering for the European Fast-Moving Consumer Goods Industry

Cathy Barnes, Tom Childs, and Stephen Lillford

CONTENTS

9.1 Introduction

During the last 15 years the number of fast-moving consumer goods (FMCG) available in the average European supermarket has grown exponentially. To be successful in this dynamic environment, companies must constantly launch many new products and build successful brands.

To support the speed of innovation and brand creation, the industry sector has many existing methods and tools incorporated within its product development processes. Thus, to be widely adopted in the FMCG arena, Kansei/affective engineering must fit within the industry's current high-speed product development processes and demonstrate that it addresses gaps in the existing suite of methods. Perhaps even more important, Kansei/affective engineering must ensure it takes full account of the needs of the product brand.

9.1.1 Building a Successful Brand

One of the distinctive factors of the FMCG industry is the importance of the brand. The brand was originally created to reassure the consumer of a quality product by building awareness and developing positive associations. Today, the brand is of significant commercial importance and stretches far beyond having a recognizable logo on the product. Figure 9.1, based on

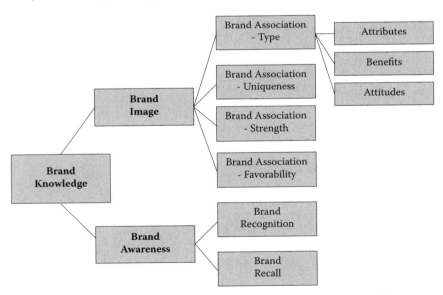

FIGURE 9.1
Dimensions of brand knowledge. (Adapted from Keller, K. L. 1993, *Journal of Marketing* 57 [January], 1–22.)

Keller (1993), shows the many different aspects that comprise the nebulous "brand" concept.

9.1.2 New Product Development in FMCG

Typically most FMCG companies use a variant of the stage-gate process. Figure 9.2 shows a reduced set of stages, from targeting a need to deploying in the market, based on a survey of companies as part of a European Commission (EC)-funded coordination action, the ENGAGE (design for emotion) project (Childs et al., 2006). The impression of a linear activity that it gives is, however, misleading. Although failure to pass a gate (between stages) will result in canceling the development, learning from a later stage can cause an earlier stage to be revisited. The "learning by doing" view of Figure 9.3 shows the stages more truly as activities that can be revisited at

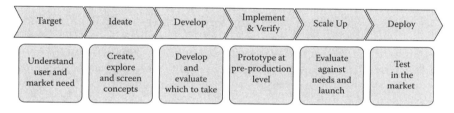

FIGURE 9.2
Steps of a stage-gate development process.

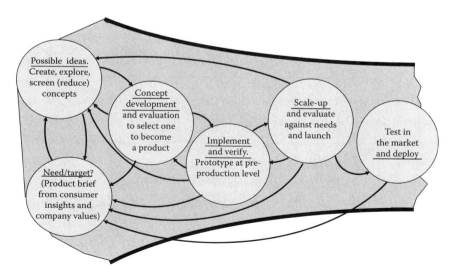

FIGURE 9.3
Learning by doing in a product development funnel.

any time, although that is more common the earlier the stage of development. Particularly at the start of a development the need is to quickly explore and screen out a large number of possible ideas and to move on. New insights during concept development and even during preproduction certainly can lead to design changes. Feedbacks from the later to intermediate stages more commonly relate to aspects such as product color or details of labels rather than, for example, shape changes.

Feedback from the market to the need is usually considered to be part of continuous improvement, signaling launch of a new development, rather than modification to an existing project. One aspect of development that is not shown by either Figure 9.2 or Figure 9.3 is who is involved. Different companies bring together teams of marketeers, industrial designers, engineers, and sales staff in different ways at different stages of the process. In some cases a multidisciplinary group is formed to take a product forward from need to deployment. In others the product is passed from group to group as it develops. In all cases there is a need for good interdisciplinary communication tools.

Designing products and packaging for consumers is considered by most businesses to be a "consumer journey," which is a chronological map of the consumer/brand interactions (Davis and Longoria, 2003). Figure 9.4 shows that there are, in the main, three *touchpoints* on the journey, and lists some of the ways by which the consumer experiences a branded product. However, two of the touchpoints are particularly important to a consumer-centered design process such as Kansei/affective engineering. These are the Purchase Experience and the Use Experience, often colloquially called the First Moment of Truth and the Second Moment of Truth, after a widely reported Procter & Gamble process. The Purchase Experience is particularly important because it occurs within the retail outlet. There the product needs to quickly and effectively communicate its benefits. The messages here impact consumer expectations, but primarily a good design for this touchpoint will certainly

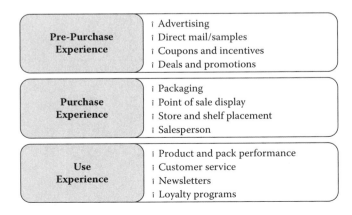

FIGURE 9.4
Experiences on the consumer journey.

sell more products in the first instance. The consumer experience at the Use Experience determines whether the product will be repurchased. A well-designed product for this stage will build brand loyalty and ensure long-term profitability.

9.1.3 Why Does FMCG Need Kansei/Affective Engineering?

It is clear that good design that appeals to consumers throughout the product's life cycle is critical to FMCG. However, it is not just the product that needs to be considered. The impact of the brand communication is critical, as is the packaging that is used to transport and contain the product. Thus, simple but effective tools are needed for companies to be able to quickly understand whether products in development will satisfy consumers and thus be successful.

Kansei/affective engineering fills a gap in the FMCG manager's toolbox. The Kansei/affective engineering variant presented in this chapter can consider all the touchpoints throughout the consumer journey and measure quantitatively whether a concept matches consumer needs. It can also ensure that the brand equity is incorporated within the design and can also test the effectiveness of that communication, critical in this industry. This Kansei/affective engineering is also of benefit over and above more traditional design and market research techniques as it can be used to create a set of assessments from untrained consumers rather than from a trained panel of people. This can both reduce the costs of testing and ensure a wider population sample can be used in the research. A Kansei/affective engineering system, as reported earlier in this book, is rarely needed in the FMCG industry, as the test samples used in the research are not the final shelf-ready design but rather will form a set of fundamental guidelines for the solution. Finally, Kansei/affective engineering provides a unique tool by which all the disciplines involved in product development can be brought together in a team to use a single process.

9.2 Kansei/Affective Engineering Framework for FMCG Products

The world of FMCG is very fast moving and consumer focused and so requires specific tools to ensure these are considered throughout the development process. Figure 9.5 shows the five-stage framework of our Kansei/affective engineering methodology for FMCG products. It has developed from experience gained through some 30 sponsored projects, most of which have been carried out under confidentiality agreements. It is described step by step in this section and illustrated by elements from three case studies that we can talk about. One is based upon a project completed with a drinks

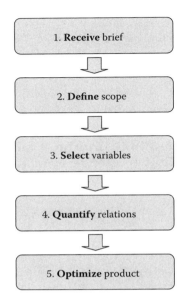

FIGURE 9.5
Kansei engineering framework for FMCG products.

company that had introduced a new bottled product to the UK market. It was an extension to a well-developed brand. The aim of the study was to explore how well the launched product communicated the brand's values as well as the product's intended unique benefits, and to recommend any design changes that would improve user perceptions. The second relates to a project on a personal skin-care product, namely, the container for a body moisturizer. The project was less constrained than the first. Although the product too was to be a branded one, it had not at the time been released on the market, although concept work had started. Guidelines relating to shape, color, and texture (for touch feeling) were requested. The third relates to confectionery product packaging. It was the least constrained of all. Again, it concerned a branded product, but only a need had been identified. The company was open to all ideas about what the product's form should be. These three examples can be seen to fit into different areas of the development funnel (Figure 9.3).

The methodology has five stages to support the designer:

Stage 1: *Receive* brief
Stage 2: *Define* scope
Stage 3: *Select* variables
Stage 4: *Quantify* relationships
Stage 5: *Optimize* product

9.2.1 Stage 1: Receive Brief

The process starts from a high-level design brief concerning the need to be met, including target audience and imposed brand guidelines. In the case of the bottled drink project, the brief was for a drink for young males who enjoy a short social drink with close friends and prefer premium alcoholic drinks. For the confectionery project, the brief was to create packaging to support sharable product moments, and a target group was specified. The moisturizer project was positioned slightly differently. Its brief was to explore what attributes of the container were most important to women who were willing to pay more for products that were good for their skin and who felt neutrally about or were dissatisfied with their existing lotion. Out of this, the guidelines for shape, color, and texture, referred to in this section's introduction, emerged.

9.2.2 Stage 2: Define Scope

The next process stage is to determine exactly what is to be the scope of the investigation, by defining its context and generating the experimental variables. It could be argued that defining the context is a detail of the first stage. In this respect the division into stages is artificial. A continuous development is involved. Generating the experimental variables is, however, certainly separate from the first stage. The experimental variables are the independent and dependent variables of product attributes and evaluation descriptors that are the common aspects of Kansei/affective engineering activities. At this stage the object is exhaustively to generate variables, many more than can be followed up in detail. The purpose is as completely as possible to consider the design space. Reduction of variables occurs in Stage 3.

9.2.2.1 Context

First, which touchpoint is to be investigated needs to be defined. It is possible to study more than one touchpoint in a study, but this requires separate experiments to be run for each. Consumer interactions with products are highly context specific. The most relevant location and environmental conditions must be specified along with determining the exact type of interaction with the product.

Taking the bottled drinks product as an example, many aspects contribute to the whole experience of it: the flavor, the bottle design, and the advertising. The flavor is developed through sensory exercises with trained panels using defined descriptors like *sweet*, *crisp*, and so forth. Advertising is outside the scope of this chapter. The case study was concerned with the bottle design. To select an appropriate design, what qualities are communicated through different bottle designs, what attributes of bottles contribute to communicating different qualities, and what might be done to communicate the ideal qualities more effectively must be clarified. These aspects require the responses of the specific target users to be understood rather than those of a skilled, trained panel. It is here that the sponsoring company found Kansei/affective engineering to give it insights it did not already have. Several points of interaction between bottle and user could have been explored: viewing the bottle on the bar shelf (the product's social context), holding the bottle, opening the bottle, drinking from the bottle, and disposing of it. The scope of the case study was purely on-the-shelf appearance, the purchase experience of Figure 9.4.

9.2.2.2 Evaluation Variables

As with all versions of Kansei/affective engineering, a set of variables needs to be defined that will become the criteria against which all the product concepts will be evaluated by consumers. Often they are word variables, usually adjectives, but they can be phrases. They can also be nonverbal. Hereafter, the terms *evaluations*, *adjectives*, or just *words* will be used interchangeably as seems best to fit the context. Words must be included that describe the full range of desirable product qualities.

Target users' words are collected from a series of discussion exercises in which a recruited target group is asked to interact normally with a set of relevant products and related materials (e.g., images). Their preferences are interrogated using structured exercises such as

1. Articulation of product preferences and reasons
2. Sorting products into groups by both similarities and differences, similar to Kelly's repertory grid method (Van Kleef et al., 2005)
3. Describing ideal or ultimate products

Figure 9.6 shows views of a focus group discussion, both with products as screen images and a wide range of collected actual products, taken during the body moisturizer case study. Words also need to be chosen that accurately describe the essence or personality of the brand. Without them, judgments relating to the brand image will be lost from the experiment. We have found that words supplied directly from the brand owners' resources of brand and product descriptors are often not suitable. Some problems that can arise are as follows:

- Users may not understand a word, resulting in an incorrect response.
- The words may have similar meanings to other words; this can cause results to incorrectly weight a particular response.
- Users can be led or confused by unfamiliar words or by words that may be difficult to consider in the required context (e.g., Is this wrist-watch *oppressive*?).

FIGURE 9.6
A focus group session.

We have developed a way to avoid such problems. Words supplied by the brand owner are used as seeds to grow more words that are their synonyms, using the British National Corpus of language (Delin et al., 2007). At Stage 3 (Section 9.2.3) it is decided which of them, with the target user–generated words, to use. More detail of the Corpus tool is presented in Section 9.2.3.

Returning to the bottled drinks product example, small focus groups of 5 to 10 target users were held. Structured exercises were completed in the areas of articulating reasons for preferences (e.g., "Describe your preferred drink and why you like it.") and stimulating insights using previously developed mood boards (see Design Attribute Variables, below). Brand and product benefit adjectives were collected and expanded using the Corpus tool. A selection of the many hundreds of resulting words is shown in Table 9.1.

9.2.2.3 Design Attribute Variables

Similarly to the collection of evaluation variables, a large number of products/ product concepts is initially explored from which to generate design attribute variables that might be expected to contribute to the overall affective

TABLE 9.1

Some of the Valuation Words from the Bottled Drink Case Study

Product and Flavor Qualities			Brand Essence		
Lime	Crisp	Refreshing	Masculine	Modern	Urban
Citrus	Crisp	Arousing	Angular	Advanced	Bustling
Fresh	Dry	Energizing	Bold	Contemporary	City
Lively	Edge	Exhilarating	Definite	Current	Dark
Natural	Firm	Invigorating	Heroic	Influential	Downtown
Piquant	Fresh	Keen	Intense	Hip	Ghetto
Juicy	Piquant	Lift	Macho	Latest	Industrial
Zesty	Ripe	Refreshing	Manly	Modern	Metropolitan
…	Sharp	Regain	Masculine	Now	Neon
	…	Regenerate	Mature	Popular	Suburban
		Renew	Muscular	Powerful	Urban
		Rejuvenating	Potent	Sophisticated	…
		Reviving	Powerful	Successful	
		Rousing	Robust	Trendy	
		Stimulating	Strong	…	
		Tonic	Vigorous		
		…	…		

quality of the product. The generation of sample products from which to extract affective attributes by experiment is straightforward when a large range of products similar to the target product already exists. Sources are shops, catalogs, magazine photographs, and company (including competitor company) Web sites.

Figure 9.6 has shown an example of collected products that were used to stimulate the generation of evaluation variables in the moisturizer case study. These were also used to analyze the typical shapes of moisturizer bottles. From these, images of 48 shape samples were created: jars, cylinders, flasks, downward and upward tapering, waisted, oval, teardrop, and tubular forms. Some were later made as three-dimensional objects by rapid prototyping. They are shown in Figure 9.7. In addition, 72 color patches were generated for color tests. Forty-nine plastic, fabric, paper and board, metal, and miscellaneous samples were collected for tactile tests.

The position is more complicated the more one has to rely on the generation of novel concepts as part of the project, that is, for radical new products. The position is most problematic of all when trying to develop new product attributes to relate to brand qualities, such as *heroic*. Such descriptors seem to be too subjective. Our investigations suggest that transferring brand quality traits to user perceptions of products is particularly difficult.

To overcome this we have developed a new technique based upon mood boards (McDonagh et al., 2002), which has the added advantage that it is aligned with existing practice. Mood boards are common design tools created by a multidisciplinary team and use images, colors, forms, and textures

FIGURE 9.7
A selection of rapid prototyped moisturizer container shapes.

to communicate the feel of a conceptual product to others. A set of images is chosen by the design team to cue the brand qualities. If appropriate, these are checked with consumers to ensure suitable images are used. These images are then used as stimuli in a multidisciplinary brainstorming study to translate them to product attributes. Figure 9.8 is a montage sample from the confectionery product case study. Typically, many tens of concepts are generated, and these are sorted to reveal around 20 diverse concepts. The criterion for selection is team consensus. They would be added to products collected more straightforwardly from the existing market.

9.2.3 Stage 3: Select Variables

This stage determines which adjectives and product concepts are used in the main experiment. Kansei engineering typically uses a large number of both adjectives and concepts. This makes for a very long and costly experiment, which does not fit well into current FMCG product development processes. It is essential to find robust ways to reduce the number of adjectives and concepts.

Sharable product moments

Exciting   *Friendly*

Generous *Family fun* *Togetherness a treat*

Informal

Surprise

Packaging concepts

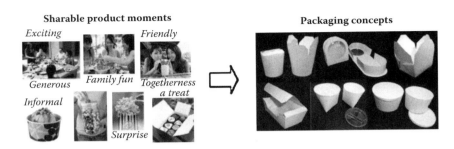

FIGURE 9.8
Mood boards and process of translation to concepts.

9.2.3.1 Evaluation Variables

The outcome from Stage 2 is typically several hundred collected adjectives, from both the target users and brand owners, describing perceived product qualities. The total from both groups must be reduced to an overall total of around 15 to 20, with some balance kept between them. We have developed two procedures to support the reduction process. One of the two is a set of rules involving a simple set of linguistic guidelines. Table 9.2 lists a selection of them. Words from both groups are assessed against these rules to determine whether they will be easily understood in the context of the experiment. Unsuitable words are eliminated from future consideration.

The other procedure relates specifically to the brand-generated words. Table 9.3 was generated, based on the moisturizer study, to help describe the

TABLE 9.2

A Selection of Five Rules from a Total of Almost 20 for Rejecting Evaluations

Rule 1	Remove evaluations that are not plausibly related to objects	While "friendly," for example, may be metaphorically extended to apply to inanimates, words such as "enthusiastic" and "unbiased" may not be.
Rule 2	Remove evaluations that describe purely evaluative reactions	For example "good" and "nice" do not describe products but people's feelings about them. Feelings of like and dislike do not inform us about specific product-related reactions.
Rule 3	Remove ambiguous evaluations	For example "clear" might be a good word to refer to either a concrete or an abstract quality of products that promote cleanliness (i.e., clear skin) or that are themselves without color but it is also frequently used to describe the clarity of an argument or conclusion.
Rule 4	Remove non-gradable evaluations	For example "unique" is not gradable, something cannot be "very unique" or "slightly unique." So using it for a semantic differential questionnaire could be ineffective.
Rule 5	Remove evaluations that are out of context with the research	Remove adjectives that relate to untested senses or that relate to a prolonged experience with the object rather than the controlled conditions of the Kansei study. For example, "satisfying" for an appearance.

TABLE 9.3

Translating Brand Words to User Words with the Corpus Tool

		Love	Delight	Happy	Comfortable	Believable	⋮	⋮	Skin kindness	Smoothness	Balanced	Moisturize	Clarifying	⋮	⋮	Number of seed words synonyms relate to
	Number of synonyms the seed word relates to	8	4	3	7	4	8	6	3	3	4	6	2	3	7	4
Synonyms from Corpus Tool	Tender	✓			✓		✓		✓	✓		✓				6
	Conventional				✓	✓	✓	✓				✓			✓	6
	Fun	✓	✓	✓				✓							✓	5
	Luxurious	✓	✓		✓		✓				✓					5
	Showy		✓			✓	✓	✓					✓			5
	Everyday				✓	✓		✓			✓			✓		5
	Slender				✓		✓			✓	✓				✓	5
	Cozy	✓			✓		✓			✓						4
	Bold				✓		✓						✓	✓		4
	Friendly	✓		✓			✓								✓	4
	Simple					✓			✓		✓				✓	4
	Traditional				✓	✓		✓							✓	4
	Natural							✓			✓				✓	3
	Romantic	✓	✓				✓									3
	Advanced										✓			✓	✓	3
	Amusing	✓		✓												2
	Spiritual	✓					✓									2
	Casual								✓					✓		2

TABLE 9.4

The Final Word Selection for the Bottled Drink Study

Intense	Powerful	Sophisticated	Crisp	Lively
Rough	Refreshing	Masculine	Urban	Modern

procedure. Its left-hand column contains the seed words (Section 9.2.2) from the brand owner. Its top row holds some of the synonyms generated from the Corpus tool introduced in Section 9.2.2. These words are placed from left to right in the order of their importance as judged by how many of the seed words they relate to. For example, the word *tender* relates to six of the seed words, as indicated by the ticks in the table elements and totaled in the bottom row. How many times the seed words are related to the synonyms is also totaled along the rows of the matrix and entered in the right-hand column. Synonyms are selected for the main experiment on the basis of their relevance (lower row scores) and inclusiveness (ensuring that seed words remain represented), up to the maximum number that is judged appropriate for the experiment considering that target user words must also be chosen. Normally they would be found from among the 10 top ranked in terms of their relevance.

Usually there are too many target user words remaining after reduction by the linguistic rules. These may be reduced by thinning clusters, in the same way as can be used for reducing samples, as considered next under Design Attributes. If it is found impossible to reduce the number of words to the 15 to 20 range, it is an indication that the planned experiment is too ambitious. Table 9.4 has the final words used in the bottled drinks study.

9.2.3.2 Design Attributes

The last aspect of the experiment that needs to be defined is which design attributes should be considered and how to vary them appropriately. We use two methods for making these choices:

9.2.3.2.1 Semantic Mapping

This method is suitable when it is easy to define just two key evaluations for a product. These are chosen as the axes of a two-dimensional map that physically is a large horizontal board. A small number of consumers, 5 to 10, place the large number of product concepts on the map. Figure 9.9 shows a semantic map being created for the moisturizer shape experiment. Each numbered counter represents one sample, typically the counter's cluster. Analyzing the clustering enables the key design attributes to be identified. Any suitable technique may be used (Bech-Larsen and Nielsen, 1999); however, the Repertory Grid technique has yielded good results. Sample reduction is also based on thinning out the clusters.

FIGURE 9.9
A semantic mapping experiment in progress.

9.2.3.2.2 *Pilot Semantic Differential Survey*

This is used when it is not possible to identify just two key evaluations to test. A small number of users, around 10, are asked to rate many concepts against a small number of key evaluations on a 7-point Likert scale. Figure 9.10 shows the results of a principal component analysis for the same samples as were used for the semantic mapping, Figure 9.9. The significant design attributes of these concepts can then be extracted using any suitable technique, as with the semantic mapping method.

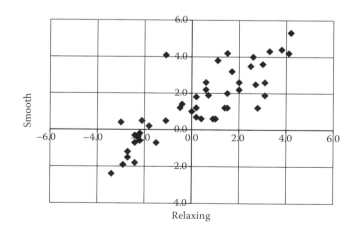

FIGURE 9.10
Outcome of the moisturizer pilot semantic differential test, with 48 samples in place.

TABLE 9.5

The Factorial Design for the Bottled Drink Study

Design Attribute	Shoulder			Body Shape		Neck Length		Color		Surface Detail			Body Width	
Attribute Style	Round	Angled	None	Straight	Curvy	Short	Long	Brown	Clear	Embossed	None	Other	Thin	Wide
1	1	0	0	1	0	0	1	1	0	0	1	0	1	0
2	1	0	0	0	1	0	1	1	0	0	1	0	1	0
3	1	0	0	1	0	1	0	1	0	0	1	0	0	1
4	1	0	0	1	0	0	1	0	1	0	1	1	1	0
5	1	0	0	0	1	0	1	1	0	1	0	0	1	0
6	1	0	0	1	0	0	1	1	0	0	1	1	1	0
7	0	0	1	1	0	0	1	1	0	0	1	0	1	0
8	1	0	0	1	0	0	1	0	1	1	0	1	0	1
9	1	0	0	0	1	0	1	0	1	1	0	1	1	0
10	1	0	0	1	0	0	1	1	0	0	1	0	0	1
11	0	1	0	0	1	0	1	0	1	1	0	0	1	0
12	0	1	0	1	0	1	0	1	0	1	0	0	1	0
13	0	0	1	1	0	0	1	1	0	0	1	0	1	0
14	0	1	0	1	0	0	1	0	1	0	1	0	1	0
15	0	0	1	0	1	0	1	1	0	0	1	0	0	1

(Sample)

After sample reduction, by thinning clusters and/or ignoring areas of the design space that might not be of interest, a factorial experimental design is constructed to test the affect of the selected attributes. Table 9.5 shows the bottled drink study design. Fifteen bottle samples were required to represent the total of 14 attributes (three shoulder types, two body shapes, two neck lengths, two colors, three surface details, and two body widths) to be taken forward to Stage 4. Among the samples there were at least two cases of each attribute style.

9.2.4 Stage 4: Quantify Relationships

This stage comprises a standard semantic differential survey with some supplementary demographic and overall liking/preference questions that allow for additional correlations and consumer groupings. Samples are rated in terms of the adjectives on a 7-point scale, using a semantic questionnaire. The order of the adjectives is randomized in the questionnaire to reduce conditioned responses and any systematic effects of participants' fatigue. An overall liking score is also collected on a separate 7-point scale.

The aggregated data from a sample of users representing the target market are used to identify optimal designs, but before results are interpreted the response distributions are examined to judge the level of consensus. If an evaluation result from a group of users shows a double peak or random distribution, then it is removed and analyzed separately. This is because these distributions could be an indication of disagreement in the meaning/use of the evaluation or that it is not suitable for relation to the concepts (despite the various linguistic tests and rules that have been applied to avoid this).

Outlier users are also identified from their responses: These are users who have scored outside two standard deviations from the aggregated response. Their responses are removed too and analyzed separately.

The cleaned data set is used to display the profile of each of the concepts against all of the adjectives. To reduce the complexity of the results, principal component analysis is used to identify which adjectives have been perceived similarly in the exercise. Those with similar meanings are given a new cluster heading, by making a judgment about the common meaning among them. The concept set can then be displayed in the principal component space to make easy comparisons.

As aggregate scores have been used for the points in the profiles and PCA, a Kruskal–Wallis test (Kruskal and Wallis, 1952) for significant differences is carried out to see if differences between scores of different concepts are important or whether there are overlaps between users. Multiple regression analysis is performed as described in earlier chapters of this book to identify and rank which attributes are contributing to which affect.

9.2.4.1 The Bottled Drink Case Study

Thirty-seven users matching the target consumer profile took part in a semantic differential survey to rate the 15 bottles against the 10 evaluations. The data were collated and outliers were removed. In this case two users were removed. A principal component analysis of the data showed three significant independent components. However, the first two accounted for more than 77% of the variance so they will be focused on here. Based on the adjectives' loading on each component, these were called *flavor intensity* and *masculine strength*. Figure 9.11a shows the type of bottle that made up the study. Figure 9.11b shows how the 15 bottles of the study loaded in the PC1/PC2 principal component space. It is seen that bottles 15 and 11 communicated flavor intensity most; 3 and 10 communicated masculine strength most; and 2, 5, and 8 communicate both qualities to some extent.

There were several bottles that did not have significantly different adjective scores when considering the raw rather than aggregated data. These have been circled in the principal component space to show that, although they locate differently, this is not significant.

A regression analysis between the attributes and the adjectives aligned with the principal components showed that the design of the shoulder shape was the most important attribute for communicating *flavor intensity* and that angled shoulders scored highest. Shoulder shape was also the most

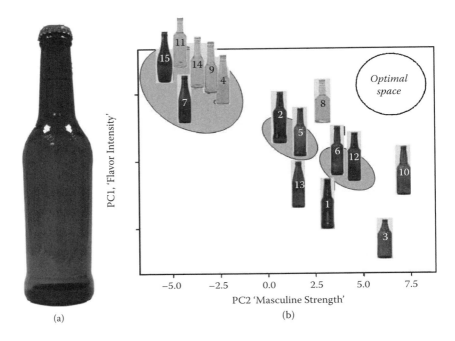

(a)   (b)

FIGURE 9.11
(a) A standard drink bottle product, and (b) a range organized by Kansei engineering methods.

important attribute for communicating *masculine strength*; however, round shoulders scored highest.

9.2.4.2 The Moisturizer Case Study

Sixty-five women took part in each of three semantic differential experiments, one each for the shape, color, and texture aspects of the design attributes. Figure 9.12a is a partial composite view of the loadings of the samples in the PC space. Further detail is in Childs et al. (2006).

9.2.5 Stage 5: Optimize Product

9.2.5.1 Target Users

Cluster analysis can be used on the overall liking scores to determine if the preferences of the participants in the experiment were homogenous or not. For example, the dendrogram of Figure 9.13 shows that in the bottled drinks case study there were three distinct participant groups within the sample. There are two large groups: one of 25 users and the other of 8 users, and one small group of two users. The scores showed that the group of 8 liked wide bottles whereas the group of 25 did not. The other two users liked both. This result suggests not a single but two possible optimal design solutions.

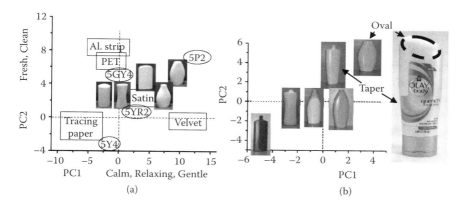

FIGURE 9.12
(a) Composite result of moisturizer shape, color, and texture experiments; (b) combined validation experiment.

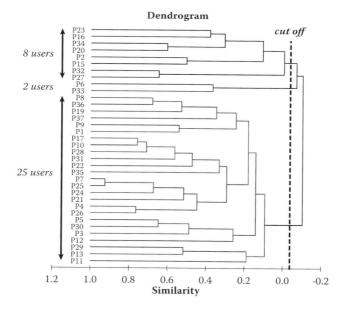

FIGURE 9.13
A dendrogram from the bottled drinks case study.

9.2.5.2 Design Space

To correctly interpret how the principal component (PC) space can be used as a visual design guide, the construction of the space must be checked carefully to determine whether evaluations are correlating negatively or positively with the space's axes. If all evaluations correlate positively and all are equally desirable to the project owner, then those concepts located highly on all the relevant PCs should be recommended. If there is no concept in this location, then a new design could be proposed that is developed from comparing concepts that locate highly on individual components. Integrating consideration of liking score with PC space provides an insight into which qualities relate highly with user preference. Concept liking scores can be overlaid or correlated with PCs but should not be included in the PC analysis itself as they have a different type of semantic meaning.

Continuing with the example from the bottled drink study, Figure 9.11, supposing the aim was to generate a bottle to support both intense flavor and strong masculinity, the optimal concept location in the PC space is indicated as in the top right corner of Figure 9.11b. The location of the existing bottles shows that the wide clear bottle (8) is the most suitable from the set, with the initial concept bottle (5) second. From the locations of the set and visual differences between the bottles we could speculate that change in the initial concept glass color and width might increase affective communication of the product.

9.2.5.3 Design Guidelines

Through a combination of both PC multivariate and regression analyses, design guidelines can be created. Many cases will require the designer to make compromises between the optimal concepts for each PC to achieve a best overall design. The regression analysis gives an order of attribute significance to inform this decision making or can provide guidance for further attributes to test that have not been included.

Again from the bottled drink case study, the regression results that show different shoulder designs are responsible for transmitting *flavor intensity* and *masculine strength* explain why there is not a bottle in the top right-hand corner of the PC space. Therefore to increase communication of both PCs beyond the potential of the variables tested requires a new design of shoulder shape or a new attribute to be tested. From the range of attributes tested we are able to speculate that the shape of bottle 5 communicates the combined qualities fairly well, and a possible alternative would be a wide, clean bottle (based on bottle 8).

In the moisturizer case study, a verification experiment was carried out, combining the shapes and colors expected to create a most clean and gentle effect. Figure 9.12b shows the result. In this case the experiment's samples were simpler than the finally launched product, inset in the figure. However, its oval and downward tapering shapes and heather coloring (code 5P2 in Figure 9.12a) are reflected in the final product. It helped confirm the product development direction.

9.3 Conclusion

Within the FMCG industry the Kansei/affective engineering methods that are the subject of this chapter are widely seen by new product development teams to offer a sufficient number of benefits to make them useful. They are seen as not standing alone but fitting in with and filling gaps between other new product development tools. In particular, they are not used to provide complete design solutions but to define key parameters of a solution to be fleshed out in more detail by other means. In an industry that frequently relies on expert panels for opinions, they also provide a way to involve untrained users (closer to the general consumer) in the development process. They also generate rational reasons for decision making that all the different disciplines involved in new product development can respond to. This communication role lasts after a product's launch, through archival documentation that can be referenced for future developments.

However, Kansei/affective engineering is not universally accepted. It can generate confusion and skepticism. It is perceived by some designers as conflicting with or restricting their own skills and intuition. It is also perceived by some designers and innovation managers as constrictive to the design process itself. By bringing untrained users into the development process, seen by supporters as one of its advantages, it certainly results in decision making relying less on the flair of individuals. The more Kansei success stories can be told, for example, through books such as this, the more the negative perceptions may be countered. However, what may be a best way to balance individual flair and group assessment within a structured process is an area that could benefit from further development.

Acknowledgment

The work reported in this chapter includes work carried out as part of Knowledge Transfer Partnership No 6179, to "apply Kansei Engineering to the European packaging industry" in conjunction with Faraday Packaging Partnership and PIRA International, Ltd. The authors of this chapter would also like to thank all the participating company members of the Faraday Packaging Partnership who had significant input into the definition of this process. Special thanks must go to the client companies who provided the case studies reported here.

References

Bech-Larsen, T., and Nielsen, N. A. (1999). A comparison of five elicitation techniques for elicitation of attributes of low involvement products. *Journal of Economic Psychology*, 20, 315–341.

Childs, T. H. C., Agouridas, V., Barnes, C. J., and Henson, B. (2006). Controlled appeal product design: A life cycle role for affective (Kansei) engineering. In *Proc. 13th Int. Conf. Life Cycle Engineering*, Leuven, Belgium, pp. 537–542.

Davis, S., and Longoria, T. (2003). Harmonising your touchpoints. *Brand Packaging*, Jan/Feb.

Delin, J., Sharoff, S., Lillford, S., and Barnes, C. (2007). Linguistic support for concept selection decisions. *AI EDAM* 21, 123–135.

Keller, K. L. (1993). Conceptualizing, measuring, and managing consumer based brand equity. *Journal of Marketing* 57 (January), 1–22.

Kruskal, W. H., and Wallis, W. A. (1952). Use of ranks in one-criterion variance analysis. *Journal of the American Statistical Association* 47, 583–621.

McDonagh, D., Bruseberg, A., and Haslam, C. (2002). Visual product evaluation: exploring users' emotional relationships with products. *Applied Ergonomics*, 33, 231–240.

Van Kleef, E., van Trijp, H. C. M., and Luning, P. (2005). Consumer research in the early stages of new product development: A critical review of methods and techniques. *Food Quality and Preference*, 16, 181–201.

10

Kansei/Affective Engineering Applied to Triggers in Powered Hand Tools

Ebru Ayas, Jörgen Eklund, and Shigekazu Ishihara

CONTENTS

10.1 Introduction

Designing products for highly competitive markets requires thorough consideration of the preferences of customers and users. Kansei/affective engineering is a well-known approach that considers the emotional responses in the design of products and services (Nagamachi, 1997). A majority of the studies published in scientific articles deal with appearance and function of the products at stake and consequently relate to perception. However, engineering design also deals with issues that have consequences for other dimensions of human perception. Technical solutions may, for example, be engineered to produce specific sounds, as the sound of closing a car door, the sound of a vacuum cleaner, or the sound of an accelerating car engine (Nagamachi, 1997). In other cases the olfactory sense is addressed (Schifferstein and Hekkert, 2007; Nagamachi, 2001), for example, when designing the particular smell of the interior of a new car. The tactile feeling when using products is another field in which interest is growing. Examples of this are the mechanical properties of the foot platform of a lift truck (Axelsson et al., 2001) and the properties of rocker switches in vehicles (Schutte and Eklund, 2005). In these cases, strong emphasis is given to properties that the user experiences in the situation of using the product, both initially and after a long time.

Kansei/affective engineering may be used for many different dimensions of feelings and experiences. Driver feeling in vehicles is one example and safety feeling is another (Axelsson et al., 2001). Quality feeling is another example discussed increasingly often (Ayas, 2008). In the field of ergonomics, user feeling is becoming an extension of usability, following the proposed model of requirements of product design: design for function, for usability, and for emotion (Childs, 2006).

This chapter focuses on the design of switches for handheld nut runners, as this is an important aspect in ensuring good usability, feedback, control-lability, safety, and quality, which are part of the user feeling.

10.2 Trigger Design

A trigger switch functions as an interface between the operator and the hand tool in order to obtain safe and controllable operation of the hand tool. The actuation and switching mechanisms need to be designed to afford precise control of different loads of hand tools. Tactile feedback occurs in the form of a properly designed force–travel relationship (Lepore et al., 2006). A recent study shows that on average, participants could only reliably identify two to three stiffness levels in the range of 0.2–3.0 N/mm, and two to three force magnitude levels in the range of 0.1–5.0 N. It is recommended that high and low stiffness or force–magnitude levels are used, with an additional third level (medium) for more experienced users (Cholewiak et al., 2008).

The perceptual thresholds for dynamic changes in a rotary switch were evaluated in two experiments using an adaptive procedure (Yang et al., 2003). They measured humans' abilities to detect the presence of a random noise superimposed on a sinusoidal torque versus angular position. The detection thresholds were found to be in the range of 1–3% of the peak torque.

Power tool trigger forces should be high enough to avoid accidental activation, but not so high that they fatigue the fingers used to hold them down during tool use (Eastman Kodak Company, 2004). Mital and Kilbom (1992) and Eastman Kodak Company (2004) recommend that the force required to pull an index finger–activated trigger should not exceed 10 N. Moore (1975) provides recommendations {min 3 mm, max 6 mm} for displacement for push buttons for different fingers operating sequentially or randomly.

Tests have shown that poor force–travel relationship is a primary source of missed keystrokes in high-speed data entry for keyboard buttons (Lepore et al., 2006). Skilled operators develop a rhythm of depressing keys. They keep several steps ahead of their actual manipulation, counting on a change in their rhythm as a signal to stop (Lepore et al., 2006). Building "feel" into power-assisted controls presents difficulties. However, much can be achieved by devices such as the spring-loading of controls to indicate the zero or central position, and by a change in pressure as well as control position to give the operator a better knowledge of his actions (Eastman Kodak Company, 2004).

Weir et al. (2004a) tested three different linear switches, which are described as *clicky*, *smooth*, and *mushy*. The study measured dynamic quantities during human actuation of a switch. The *clicky* switch has two stable states, similar to a retractable ballpoint pen. The *smooth* switch is exactly that, a smooth momentary switch with no intentional features other than a solid feel. The *mushy* switch is also a momentary type, but with a discernable detent or over travel feel. They presented a graphical representation technique called *haptic profile*, which characterizes the feel of a switch when it is activated. They point out for further research that the combination of features that results in a pleasant feel has not been answered yet in the literature.

Schütte and Eklund (2005) identified three design factors from 29 Kansei words to evaluate rocker switches, namely *robustness*, *precision*, and *design*, which were strongly influenced by the zero position, the contact position, the form ratio, the shape, and the surface of the rocker switches. Using Quantification theory Type I, *quality* was defined as a combination of characteristics for a rocker switch: narrow form ratio, zero position to the side, no stay in contact position, and with indentures on the surface.

Kosaka et al. (1993) made sensory tests with a questionnaire using the words *initially smooth, deep, clicking, stiff, arriving shock, clear, loud sound, stiff sound*, and *sharp sound*. The study was carried out to acquire sensory data that relate both to the feeling and the switch's physical characteristics. The pattern of the degree of the touch feeling was expressed by the words used for training the neural networks, and the physical characteristics of the switches were used as output data.

The operation feeling for keyboard switches (Kosaka et al., 2005) was investigated by means of Kansei/affective engineering. Initial reaction, peak reaction, drop reaction, and final reaction forces were evaluated *by initially smooth, smooth, deep, clicking, stiff, arriving shock*, and *clear*. These words did not involve sound descriptors such as *stiff* and *loud sound* from their previous study in 1993. Dual scaling integrated with neural networks estimated that some of the relations between a reaction force and the quantified data of a word used to evaluate a touch feeling were nearly *linear*. Their results also showed that the feeling of depth is proportional to the initial reaction force and inversely proportional to the final pressure.

When a keyboard switch is pushed, one can feel a reaction force that changes depending on the depth to which the switch is pushed, that is, pushing "travel" (Kosaka et al., 1993). In general, the reaction force is nonlinear to the travel, and the nonlinearity determines the switch push feeling. When we push the switch, the reaction force reaches a maximum value, and after the maximum, it decreases suddenly. This sudden change in the reaction force gives the shock to the finger. The shock is referred to as *clicking*. After the clicking, the reaction force gradually increases and arrives at the bottom of the switch and provides a very big reaction force (Kosaka et al., 1993).

To define trigger feeling, 124 Kansei words (descriptors) were collected. These were reduced to 52 by affinity analysis and evaluated by operators and product developers using semantic differential technique. According to the results from average ratings, the Kansei word *ergonomic* was rated as the most important descriptor for trigger feeling, together with *user-friendly, easy to use, long lifetime*, and *comfortable* for both operators and product developers. Six factors were extracted—professional performance, safety and tactile feeling, usability, smooth operation, communication and durability, convenient and comfortable—to define trigger feeling based on the operator group's responses. Correspondingly, five factors—robust and appealing, ergonomics and operator performance, controllability and predictability, creativity, and modern and powerful—were distinguished for the product development group. Results showed that the start phase and especially quick start of the trigger mechanism is more important to operators, while end feedback is more important to product developers (Ayas and Eklund, 2009).

This present study aims to evaluate the tactile feeling related to four different trigger mechanisms and how the preferences differ for them. In Figure 10.1, we assume that tactile feeling feedback is found in the sense that the operator feels and differentiates the forces of F_1: initial reaction force, F_2: peak reaction force, F_3: drop reaction force, F_4: final reaction force, F5: difference between push-release forces, D_1: travel, and D_2: drop displacement. (F_5 is calculated as an average over the curve.)

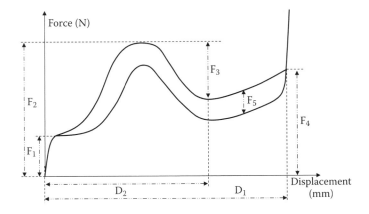

FIGURE 10.1
Operating characteristics diagram of a nonlinear trigger mechanism.

Two reference studies have been found that apply Kansei evaluation to nonlinear mechanisms of keyboard switches (Kosaka, 1993, 2005). To our knowledge, in the literature, a Kansei evaluation of nonlinear trigger mechanisms for powered hand tools and also F5 (difference between push–release forces) have not been evaluated.

10.3 Materials and Methods

10.3.1 Selecting the Trigger Mechanisms

The test rig for measuring trigger characteristics (see Figures 10.2–10.5) is built of an L-shaped steel beam and consists of a force transducer mounted on a position sensor, which allows force and displacement to be measured simultaneously. The hand tools were fixed in a stable position, and the force transducer was manually pushed against the trigger. The measurement signals from the two channels were calibrated, and data were collected in a B&K Pulse system for further treatment in Matlab™.

Four triggers were used in this study, and their physical parameter values are given in Table 10.1. The first trigger mechanism represents a nonlinear mechanism that has not been used by the operators before. It was built into two nut runners with two different installations of force–travel distance combinations. The force–travel graphics of each are presented as follows: For the two installations of these nonlinear mechanisms, namely, A (Figure 10.2) and B (Figure 10.3), a short travel distance was selected (3 mm) in A and long travel distance (5.75 mm) in B.

The next trigger mechanism (C) (Figure 10.4) also displays a relatively nonlinear mechanism, and the last trigger mechanism (D) (Figure 10.5) has a built-in linear mechanism that allows linear force increase and release. Mechanism C has a low trigger actuation force level and low travel

TABLE 10.1

Physical Parameter Installation Values for the Selected Trigger Mechanisms

Physical Parameters	Peak Force (Newton)	Initial Force (Newton)	Drop Force (Newton)	Difference Push–Release Force (Newton)	Final Reaction Force (Newton)	Travel (mm)	Drop Displacement (mm)
A	5	2	0.5	1.5	9	3	3
B	5.5	0	1.25	3	9	5.75	1.5
C	3.6	0	0.75	0.25	7	2.25	1
D	7	2	0	1.25	15	4.25	0

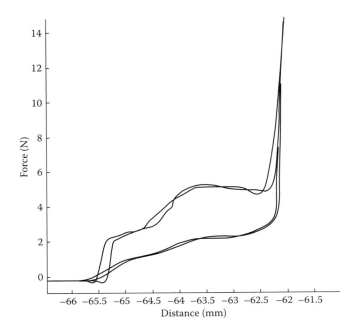

FIGURE 10.2
Nonlinear operating trigger mechanism (A).

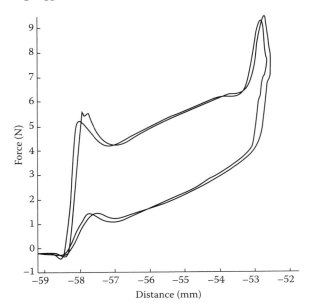

FIGURE 10.3
Nonlinear operating trigger mechanism (B).

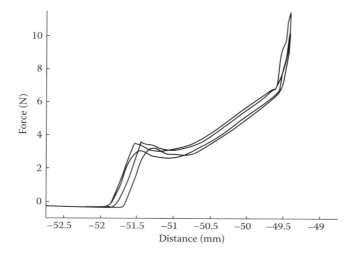

FIGURE 10.4
Nonlinear operating trigger mechanism (C).

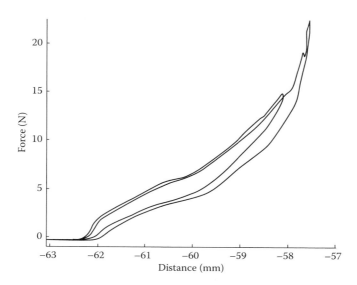

FIGURE 10.5
Linear operating trigger mechanism (D).

distance, while mechanism D has a high force level and a medium travel distance level.

The trigger feeling evaluation questionnaire consisted of two parts. The first part concentrated on understanding responses for the feeling-descriptive words and the second part was related to whether operators can distinguish between different mechanisms. In order to consider a "complete" list of descriptors that could possibly underlie trigger feeling, 124 Kansei words (descriptors) were collected from literature (databases: Science Direct, Scopus, Ergonomics Abstracts, IEEE), from interviews with engineers and product designers, and from workshops with product designers. These words were reduced to 10 (Table 10.2) by a group consisting of engineers (3), ergonomists (3), and design engineers (5) from a hand tool manufacturer. As an assisting approach, Just about Right (JAR) scales were used to assess operators when responses for hardness, maximum force, and travel distance were asked. The

TABLE 10.2

The Trigger Feeling Questionnaire Used (translated from Swedish)

Item	Questionnaire Part I (Feeling evaluation)	1–5 degree scale
1	How do you feel about the quality of the trigger?	Not good quality–Very good quality
2	Do you think that it feels good to press the trigger?	Feels not good at all–Feels very good
3	How comfortable is the trigger?	Not comfortable at all–Very comfortable
4	How safe does the trigger feel?	Not safe at all–Very safe
5	How distinct is the trigger?	Not distinct at all–Very distinct
6	How much feedback does the trigger give?	No feedback at all–Very much feedback
7	How do you experience the repeatability of the button (if you think of long time usage)?	No repeatability–Very much repeatability
8	How do you rate the tools after giving the best feeling?	Worst–Best
9	Do you think that you have control over the tool when you use the trigger?	No control at all–Very much control
10	Do you feel a difference in required force during pushing down the trigger?	Feel no difference–Feel much difference

	Questionnaire Part II (Physical parameter evaluation)	1–5 degree JAR scale
11	What do you think about the maximum force (total force that is needed to press the trigger to the bottom)?	Too low–Just about right–Too high
12	What do you think about the travel distance?	Too short–Just about right–Too long
13	What do you think about the hardness/acceleration)?	Too soft–Just about right–Too hard

operator group consisted of 23 male operators (experience 3–10 years, average experience of 6 years) working at a vehicle assembly plant in Sweden.

In a Kansei evaluation experiment, the operators were given the four tools and were encouraged to push and feel the tool characteristics. After trying out each tool they were asked to fill in the questionnaires and were free to try the tools again while answering the questions.

Partial least squares regression (PLS) (Wold, 1966) was used to predict trigger feeling. PLS is an alternative technique to ordinary least squares regression when strong collinearities between predictors are present in the model $y = X\beta + e$ (Saporta and Niang, 2006). PLS regression was performed using the SAS JMP software. Average preference values for the 10 Kansei feeling words were submitted to the PLS regression. Cross-validation was used to validate the number of factors required, that is, fitting the model to part of the data and minimizing the prediction error for the unfitted part (JMP manual, 1989–2007). The model with the lowest prediction root mean square error (RSME = 1.071) is selected. PLS is based on principal component analysis, and extracted factors are exact linear combinations of their indicators.

The Kruskal–Wallis test was used to determine whether sum of ranks for triggers are so disparate that they are not likely to have come from samples that are all drawn from the same population (Siegel, 1956). Then multiple comparison tests using post hoc Tukey were conducted to find which triggers significantly differ from one another.

10.4 Results

10.4.1 Differences in Preferences for Trigger Mechanisms

Kruskal–Wallis test results (Table 10.3) show that the trigger mechanisms A, B, C, and D differ in degree of operator's ratings for the Kansei words (p < .05).

If we take D as a reference trigger, Tukey tests in Table 10.4 show that for all Kansei words trigger D was rated significantly lower in comparison to A,

TABLE 10.3

Kruskal–Wallis One-Way Analysis of Variance by Ranks

	Quality	Feels Good to Push	Comfort	Feels Safe	Distinct	Feedback	Repeatability	Best
Chi-square	23.706	32.525	26.264	13.331	22.021	18.297	27.044	29.715
Df	3	3	3	3	3	3	3	3
Asymp. Sig.	.000	.000	.000	.004	.000	.000	.000	.000

TABLE 10.4

Multiple Comparisons Tukey Test Results for Mean Differences

Tukey HSD			Feel Good to Push	Mean Differences (I-J)					
(I)	(J)	Quality		Comfort	Feel Safe	Distinct	Feedback	Repeatability	Best
D	A	−1.30[a]	−1.69[a]	−1.56[a]	−1.08[a]	−1.73[a]	−1.08[a]	−1.69[a]	−1.78[a]
	B	−1.34[a]	−1.78[a]	−1.56[a]	−1.21[a]	−1.69[a]	−1.26[a]	−1.60[a]	−2.04[a]
	C	−1.47[a]	−1.95[a]	−1.86[a]	−1.17[a]	−1.34[a]	−1.47[a]	−1.60[a]	−2.00[a]

[a] The mean difference is significant at the 0.05 level.

B, and C (Mean = 2.43, 2.08, 2.26,; 2.65, 2.26, 2.47, 2.39, 1.82) by the operators. The highest evaluation differences were (−2.04) with B and (−2.00) with C in choosing the "best" trigger.

Mean plot of operators' responses is given in Figure 10.6. A and B were given the highest evaluation scores for *distinct* and *repeatability*, while C was given for *feels good to push, comfort*, and *repeatability*. For quality feeling all the triggers A, B, and C have similarly high evaluation scores.

10.4.2 PLS Modeling of Trigger Feeling

Table 10.5 shows the regression model coefficients predicted from the operators' responses. Cross-validation was also used to validate the number of factors required. In general, mechanisms A, B, and C contribute positively to enhance trigger feeling. The nonlinear trigger mechanisms A, B, and C contribute positively to quality feeling, while the linear trigger mechanism D contributes negatively. An increase of travel distance and of initial, peak, and final reaction forces tends to decrease the perception of quality feeling.

It should be noted that trigger A has more importance for predicting distinct feeling than the other triggers, with a coefficient of 0.17 in comparison with coefficient values of the other triggers. Trigger C does not provide a distinctive character compared to triggers A and B in the study. Moreover, drop displacement has an important effect on distinct feeling. Triggers C and D contribute more while giving feedback. For the sense of *repeatability*, trigger mechanism A has a higher effect compared with the other triggers. Considering control feeling, trigger C shows the highest effect, followed by trigger B, and trigger A shows a negative contribution to the regression model.

Wold in Umetrics (1995) considers a value of 0.8 to be a small variable importance plot. According to the variable importance values in Table 10.6, differences between push–release forces and travel distance have less importance compared with the other variables in the study. Mechanism D is also an important variable. The highest important variables were found as *final reaction force, peak reaction force*, and *drop reaction force*.

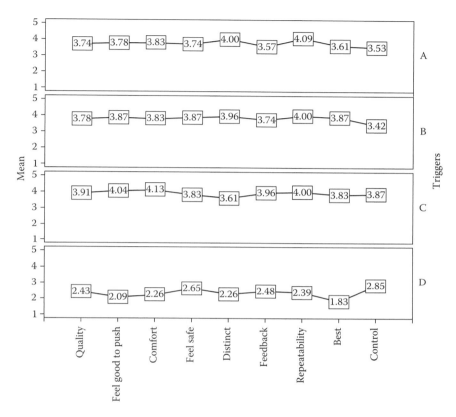

FIGURE 10.6
Mean plot of operators' responses for Kansei words for the selected triggers.

Figure 10.7 shows the PLS mapping of the trigger mechanisms and trigger parameters. Trigger A is positioned close to F5, F1, and D2; trigger B is close to D1; trigger D is close to F4 and F2; while trigger C is positioned independently from the others.

Looking at the Kansei word clustering in Figure 10.7, *distinct* and *control* are positioned as independent factors. *Repeatability* and *feedback* are also relatively distinguished as independent factors, while the rest of the words—*quality, feels good to push, comfort, feels safe,* and *best*—are grouped together.

10.4.3 Results of Physical Parameter Evaluations

For trigger A, hardness was rated as soft by 28% of the operators, as just right by 44%, and as hard by 28%. For trigger B, hardness was rated as soft by 22% of the operators, as just right by 39%, and as hard by 39%. For trigger C, hardness was rated as soft by 11% of the operators, as just right by 66%, and as hard by 22%. Hardness for trigger D was rated as soft by 17% of the

TABLE 10.5

Model Coefficients for Centered and Scaled Data

Regression Coefficients	Quality	Feel Good to Push	Comfort	Feel Safe	Distinct	Feedback	Repeatability	Best	Control
F1	-0.12	-0.13	-0.13	-0.11	-0.05	-0.15	-0.09	-0.12	-0.20
F2	-0.18	-0.18	-0.18	-0.18	-0.16	-0.19	-0.17	-0.18	-0.19
F3	0.18	0.18	0.18	0.18	0.15	0.19	0.17	0.18	0.18
F4	-0.21	-0.21	-0.21	-0.21	-0.20	-0.21	-0.21	-0.21	-0.20
F5	0.06	0.05	0.05	0.06	0.10	0.03	0.08	0.05	-0.02
D1	-0.04	-0.04	-0.04	-0.04	-0.04	-0.04	-0.04	-0.04	-0.04
D2	0.16	0.15	0.15	0.17	0.24	0.11	0.20	0.16	0.03
A	0.08	0.08	0.07	0.09	0.17	0.03	0.13	0.08	-0.05
B	0.08	0.08	0.08	0.07	0.06	0.08	0.07	0.08	0.08
C	0.07	0.07	0.07	0.06	0.00	0.10	0.03	0.07	0.15
D	-0.22	-0.22	-0.22	-0.22	-0.23	-0.21	-0.23	-0.22	-0.18

Note: F1: initial reaction force, F2: peak reaction force, F3: drop reaction force, F4: final reaction force, F5: difference between push–release forces, D1: travel distance, D2: drop displacement.

TABLE 10.6

Variable Importance Measures for the Predictors of PLS Model

Predictor	Variable Importance Plot VIP
F1 Initial reaction force	1.036
F2 Peak reaction force	1.310
F3 Drop reaction force	1.290
F4 Final reaction force	1.488
F5 Difference between push–release forces	0.326
D1 Travel distance	0.274
D2 Drop displacement	0.959
A Trigger A	0.496
B Trigger B	0.551
C Trigger C	0.659
D Trigger D	1.516

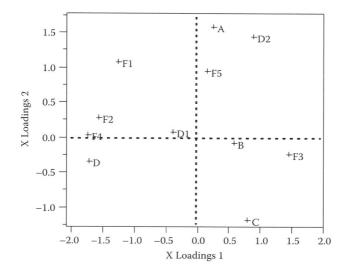

FIGURE 10.7
Mapping of trigger mechanisms and trigger parameters.

operators, as just right by 17%, and as hard by 67%. Nearly half of the operators (44%) evaluated maximum force (5 N) of trigger A as just about right and 38% evaluated it as low. Nearly half of the operators (44%) evaluated maximum force (5.5 N) of trigger B as just about right, 28% evaluated it as low, and 28% as high. Considering the maximum force (3.6 N) of trigger C, 56% of the operators rated this as just about right, and 28% of the operators evaluated the maximum force of this trigger as low. Maximum force of D was 7 N, and 11% of the operators rated this as just about right. In addition, 72% of them considered 7 N too high.

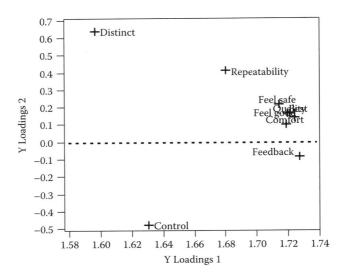

FIGURE 10.8
Clustering of Kansei words.

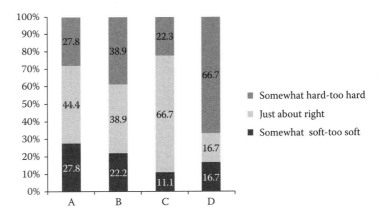

FIGURE 10.9
Hardness evaluation of triggers.

The travel distance of trigger A was rated short by 50% of the operators, long by 44%, and just about right by 5% of the operators. The travel distance of trigger B was rated short by 44% of the operators, long by 22%, and just about right by 33%. The travel distance of trigger C was rated short by 55% of the operators, just about right by 33%, and long by 11%. The travel distance of trigger D was rated short by 44% of the operators, just about right by 17%, and long by 28%.

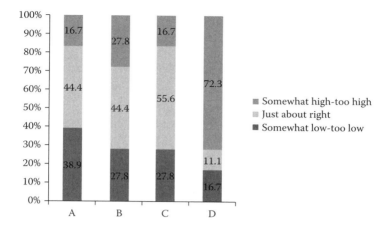

FIGURE 10.10
Maximum force evaluation of triggers.

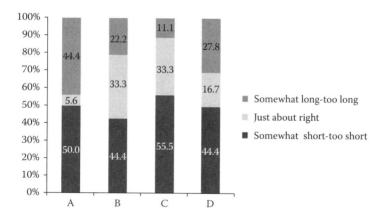

FIGURE 10.11
Travel distance evaluation of triggers.

Regarding the question whether the operators felt a force increase for trigger A, 22% rated that they felt no difference, 50% rated some difference, and 28% rated much difference. For trigger B, 11% of the operators felt no difference, 50% rated some difference, and 39% rated much difference. For trigger C, 11% of the operators felt no difference, 39% rated some difference, and 50% rated much difference. For trigger D, 27% of the operators felt no difference, 17% rated some difference, and 46% rated much difference.

Looking at the Spearman–Rho correlations of operators' responses in Table 10.7, while evaluating trigger A, their evaluation of maximum force was significantly correlated with travel distance and hardness; for triggers

TABLE 10.7

Spearman–Rho Correlations between the Physical Evaluations
of Trigger Mechanisms

Trigger	Physical	Maximum Force	Travel Distance	Hardness
A	Maximum force	–	.636[a]	.480[a]
C	Maximum force	–	.577[a]	–
B	Hardness	.768[b]	–	–
D	Hardness	.583[a]	–	–

[a] Correlation is significant at the 0.05 level (2-tailed).
[b] Correlation is significant at the 0.01 level (2-tailed).

B and D, maximum force was significantly correlated with hardness. The evaluation of trigger C for maximum force was significantly correlated with travel distance.

10.5 Discussion and Conclusions

This study on right-angle nut runners has identified design characteristics of the trigger function that are associated with different feelings and preferences by assembly workers operating nut runners in their daily work. The PLS method together with the Kansei engineering approach confirmed that operators can distinguish and show preferences for certain feelings between linear and nonlinear feedback while using a trigger. In particular, nonlinear trigger mechanisms were preferred compared with linear trigger mechanisms. Discrete on/off triggers should have a minimum operating force of 0.5 kg (5 N) and a maximum force of 1 kg (10 N) according to Fraser (1980), and progressive triggers should have an operating force toward 10 N to give precise control at intermediate control positions. The results of both approaches confirm that different nonlinear trigger mechanisms were preferred with maximum trigger forces of 3.5–5.5 N. The operators evaluated a maximum force of 7 N as very high.

The operators could also differentiate force changes in the force–displacement curve between the triggers. The responses show that operators were more positive toward the travel distance (5 mm), while the travel distance (3 mm) is perceived somewhere between too short and just right. As a complementary approach to understanding operators' tactile feelings, JAR scales were also found useful. Trigger B was preferred by the operators, and its force–distance installations were F1: initial reaction force = 0, F2: peak reaction force = 5.5 N, F3: drop reaction force = 1.25 N, F4: final reaction force =

9 N, F5: difference between push–release forces = 3 N, D1: travel distance = 5.75 mm, and D2: drop displacement = 1.25 mm.

Fraser (1980) defines that hand tools should be designed to give sensory "feedback" and be appropriate to the endurance of the user. Building "feel" into power-assisted controls presents difficulties. However, much can be achieved by devices such as the spring-loading of controls to indicate the zero or central position, and by a change in pressure as well as control position to give the operator a better knowledge of his actions (Eastman Kodak Company, 2004). From the current study results, triggers B and C are found to give more feedback than the other triggers. Looking at the parameters drop reaction force has the highest coefficient to increase feedback from the triggers. This study also has shown that the parameter difference between push and release force is positively influencing to give distinct feeling (PLS coeff. 0.10) in triggers.

Positive drivers of trigger feelings obtained in this application were *drop reaction force* and *difference between push–release forces and drop displacement*, while negative drivers were *initial reaction force, peak reaction force, final reaction force*, and *travel distance*.

When operators use tools that produce greater torque, different operation techniques and bodily motions may be involved, and these may contribute to different dimensions of their subjective evaluations (Lin and McGorry, 2009). Lindqvist (1997) and Lindqvist and Skogsberg (2007) define eight ergonomic parameters for powered hand tools on the same evaluation scale: handle design, external load, weight, temperature, shock reaction, vibration, noise, and dust and oil. We propose trigger feeling evaluation on force and travel distance sensing as a complementary measure to those parameters to design ergonomic, effective products.

More research is needed to identify in greater detail which combination of parameters gives better tactile feeling for trigger nonlinear operating mechanisms. The approach described here provides hand tool manufacturers with information for the design of operators' feelings, needs, and preferences of triggers.

Acknowledgments

The authors would like to thank everyone who supported and contributed to the study with their work, time, and expertise: Lars Elsmark, Magnus Persson, Andris Danebergs, Lars Hansson, Lars Oxelmark, Terje Ahnfeldt, Maria Savic Habo, and Ali Gorji.

References

Axelsson, J. R. C., Eklund, J., Nagamachi, M., Ishihara, S., Rydman, K., Sandin, J. (2001). Suspension and damping of a lowlifter platform—An application of Kansei engineering. In M. J. Smith and G. Salvendy (Eds.), *Systems, social and internationalisation design aspects of human-computer interaction*, Vol. 2, Lawrence Erlbaum Associates, NJ, pp. 333–337.

Ayas, E. (2008). Engineering feelings of quality, Linköping Studies in Science and Technology Thesis No.1365, ISBN 978-91-7393-898-3.

Ayas, E., Eklund, J. (2009). Identifying trigger feeling, Manuscript, Division of Ergonomics, School of Technology and Health, KTH, Sweden.

Childs, T. (2006). University of Leeds, Personal communication.

Cholewiak, S. A., Tan, H. Z., Ebert, D. S. (2008). Haptic identification of stiffness and force magnitude, *Symposium on Haptic Interfaces for Virtual Environments and Teleoperator Systems* 2008, March 13–14, Reno, NV, IEEE Computer Society, pp. 87–91.

Eastman Kodak Company (2004). Kodak's ergonomic design for people at work. *Technology & Engineering*.

Fraser, T. M. (1980). *Ergonomic principles in design of hand tools*. Occupational safety and health series. No. 44. CIS 81-1226. International Labour Office, Geneva.

JMP® (1989–2007). Version 7. SAS Institute Inc., Cary, NC.

Kosaka, H., Nishitani, H., and Watanabe, K. (2005). Estimation of reaction force of a keyboard switch based on Kansei information using neural networks, *Networking, Sensing and Control, Proceedings of IEEE*. pp. 225–230.

Kosaka, H., Serizawa, K., Watanabe, K. A. (1993). Universal keyboard switch for a feeling test. *IEEE Workshop on Robot and Human Communication '93*. Science University of Tokyo, Japan, November 3–5, pp. 225–230.

Kosaka, H., Serizawa, K., Watanabe, K. A. (1993). Universal Keyboard Switch for a Feeling Test. IEEE Workshop on Robot and Human Communication '93, Science University of Tokyo, Japan, 3–5 November, pp. 225/230.

Lepore, D. W., Williamson R. A. (2006). Switches. Pushbuttons. Keyboards. In Lipták, B. G. (Ed.), *Instrument engineers' handbook: Process control and optimization*, CRC Press, Boca Raton, FL.

Lin, J.-H., McGorry, R. W. (2009). Predicting subjective perceptions of powered tool torque reactions. *Appl. Ergonomics*, Vol. 40, pp. 47–55.

Lindqvist, B., Skogsberg, L. (2007). Power tool ergonomics, Evaluation of power tools. ISBN: 978-91-631-9900-4, Atlas Copco number 9833 1162 01.

Lindqvist, B. (1997). *Power tool ergonomics. Evaluation of power tools*. ISBN: 91-630-5217-2.

Mital, A., Kilbom. A. (1992), Design, selection and use of hand tools to alleviate trauma of the upper extremities: Part II—the scientific basis (knowledge base) for the guide. *International Journal of Industrial Ergonomics*, **10** (1–2), 7–21.

Moore. T. G. (1975). Industrial push-buttons, *Applied Ergonomics*, **6** (1), 33–38.

Nagamachi, M. (1997). Kansei engineering and comfort, *International Journal of Industrial Ergonomics*, **19** (1), 79–80.

Nagamachi, M. (2001). Thinking way of Kansei engineering and its application to a cosmetic product development, *Fragrance Journal*, **4**, 19–28.

Saporta, G., Niang, N. (2006). Correspondence analysis and classification. In M. Greenacre and J. Blasius (Eds.), *Multiple correspondence analysis and related methods*. Chapman & Hall, Boca-Raton, FL.

Schifferstein, H. N. J., Hekkert, P. (Eds). (2007). *Product experience*, Elsevier, Amsterdam.

Schütte, S., Eklund J. (2005). Design of rocker switches for work-vehicles: An application of Kansei engineering. *Applied Ergonomics, 36* (5), 557–567.

Siegel, S. (1956). *Nonparametric statistics for the behavioral sciences*. McGraw-Hill, New York.

Weir, D., Buttolo, P., Peshkin, M. J., Colgate, E., Rankin, J., Johnston, M. (2004). Switch characterization and the haptic profile, *12th Haptic Symposium on Haptic Interfaces for Virtual Environment and Teleoperator Systems*. March 27–28.

Wold, H. (1966). Estimation of principal components and related models by iterative least squares. In Krishnaiah, P. R. (Ed.), *Multivariate analysis*, 391–420. Academic Press, New York.

Yang, S., Tan, H., Buttolo, P., Johnston, M., Pizlo, Z. (2003). Thresholds for dynamic changes in a rotary switch, *Proceedings of EuroHaptics 2003*, Dublin, pp. 343–350.

11

Kansei, Quality, and Quality Function Deployment

Ricardo Hirata Okamoto

CONTENTS

11.1 Introduction

Because of the drastic change in customer needs, the trend for shorter life cycles is increasing in organizations in terms of the importance of finding product development methods and technologies that can reduce development cycle time and cost. This influences a direction to reduce the gap between true needs and expectations of the customers versus the developed product itself. One key indicator to improve is the success rate of newly developed products in the market (sales and profits). In the comparative performance assessment study regarding new products (PDMA Foundation 2004), the "best" companies have a success rate of 75.5%, while the "rest" of the companies average only a 53.8% success rate for newly introduced products. The rate of ideas that become successes in the market is 25% (1 out of 4) in the "best" companies and of 11% (1 out of 9.2) in the "rest" of the companies. The current challenges are related to improving the product development processes and reducing the gap between the final products and services versus the market true needs.

Throughout the past 70 years, the quality standards have reached such a high and competitive level that the product or service differentiation based on quality, delivery times, quality in design, production efficiency, or costs is not enough in current competitive markets. Companies and institutions have developed the ability to translate functional requirements and customer usability needs, and as a consequence, they have developed their capability to design and construct the corresponding products, services, and environments that meet these requirements. But many of these product attributes are mandatory and are already required as well as expected by the customers who do not feel fully satisfied. We are evolving from the satisfaction of the obvious and evident needs, from the functional and usability demands toward the satisfaction of profound, emotional, and affective, customer and market needs (Jordan, 2000; Green and Jordan 2002).

The translation of profound needs is a complex task because the needs are not necessarily known by the customer himself. On the other hand, the design elements of the products as well as the technical requirements to make them feasible are not necessarily known by the designers, engineers, and team of experts, so current organizations require new technologies, processes, and methodologies that allow (1) the detection of these new needs and requirements of the client, and (2) the parametric translation of these needs and requirements into the design elements, its specifications and standards, and most important, adding more value to the market.

The pursuit of this relationship between the consumer, the design, and the development of new products should be the centerpiece of a new deployment of the quality and customer satisfaction approaches. Competitive organizations are clearly oriented to the customers and have found that the translation of the customer's affective as well as emotional needs (Kansei voice of the customer) is critical in the development of a new product or service. Product quality must fit the customer's Kansei value; he/she simply wants to have an enjoyable life, and it is our responsibility to deliver the means for full satisfaction (Nagamachi 2007).

This chapter deals with the relation between the Kansei/affective engineering approaches and the new product development phases and tools, especially that known as QFD (quality function deployment), created by Dr. Joji Akao and Dr. Shigeru Mizuno in the 1960s (Akao 1994; Akao and Mizuno 1994) as a planning process to develop new products, services, processes, and technologies, as well as innovative concepts.

11.2 New Product Development Phases and Tools

New product development refers to the overall disciplined and standardized process of a company for the definition of the steps and activities to convert ideas and concepts into salable products and services. The process usually considers the concept generation, strategy setting and planning, researching, organization, resourcing, product and marketing plan creation and evaluation, as well as the commercialization of the developed individual or portfolio of products or services. Most of the developers are adopting best practices (Griffith 1997) and organization of the body of knowledge of the Product Development & Management Association (Katz 2007; PDMA 2009) which defines three macrophases of the total product life cycle as follows:

The *discovery phase* covers all the process of searching and identifying the customer's problems, needs, and benefits; defining the conceptual features, functions, and attributes to be built or created; as well as all the planning activities and the strategies to achieve these market opportunities. The discovery phase ends with the explicit definition of the formal product or service specification documents and the elaboration of the business case (plan).

The second, named the *development phase*, covers all the process of converting the product or service specifications into designs as well as the definition of all the activities to accomplish this, such as required processes, parts or components, technologies, methods, and resources. It usually includes the design, resource management, test and validation, information, and engineering technology. The development phase ends when the product or service is commercially available.

Finally, the *commercialization phase* includes the whole process of product or service production, launching into the market, postlaunching review, process improvements, performance and evaluations, management of demand, and achievement of financial goals. The commercialization phase ends when the product or service has reached the end of its life cycle and decisions are to be made regarding its retirement, renewal, or regeneration.

Different knowledge areas in the discovery phase are important in order to capture and understand all external insights from clients, buyers, users, channels, competitors, and substitute products, and anything having to do with the understanding of the voice of the customer and its translation into design and technical elements. The development phase must translate all the information captured and defined in the discovery phase into the product characteristics, as well as all the operational dimension of product innovation, processes and tools for the development and management of technical requirements, design, manufacturing, supply chain, and other dimensions for the product creation, process standardization, and improvement.

Various tools and methodologies are used along each one of the phases and also depending on the main objectives or the area of knowledge, such as customer and market research, technology and intellectual property, strategy and planning, people and teams, alliances, and finally, process, execution

TABLE 11.1

Phases and Tools in the Development of a New Product

Discovery Phase	Development Phase	Commercialization Phase
Growth share model	Design automation tools	Advertisement
Benchmarking	Design of experiments	Customer service
Business case/business plan	FMEA	ERP
Ethnography (Mariampolski 2006)	KAIZEN activities	KAIZEN activities
Competitive intelligence (Kahaner 1998)	Kansei/affective engineering	Management
		Market research
Conjoint analysis (Green et al. 1999)	Market testing	Outsourcing
Kansei/affective engineering (Nagamachi 1999, 2004)	QFD	Management systems
	Simulation	
Patent mapping and mining (Kahn 2003)	Technology road mapping (García and Bray 1997)	
Pugh analysis	Toyota production system	
QFD (Akao 1994)		
TRIZ (Altshuller 1999)		
Voice of the customer analysis (Katz 2004, Shillito 2001)		

Source: Hirata, R. (2009). Traducción de las emociones y sensaciones del cliente en productos y servicios, Ph.D. Dissertation, UNAM, Graduate School of Management Science, Mexico City, Mexico.

and metrics. Table 11.1 shows an example of the tools and techniques commonly used in each of the product development phases (Hirata, 2009).

Of all these tools, two of them cross the discovery phase and are used in the development phase articulating the market needs, definition of technical requirements, and design elements, as well as the definition of the final product or service attributes or quality characteristics. These tools are QFD and Kansei/affective engineering.

11.3 Kansei/Affective Engineering

Kansei is a Japanese word with no direct and precise translation to English or Spanish, but its meaning is nearer to a psychological feeling rather than an emotion. In the new product development context, it can be defined as the image a person has of a determined product, environment, or situation, when sensed through the senses of sight, hearing, taste, smell, and touch. Kansei is the consumer's psychological feeling and mental image regarding a product or service (Nagamachi 2004). Kansei needs are not easy to measure because they depend on the individual experience and environment, but also because their meaning depends on the context, the time, and the culture of the region or country. The meaning of *elegant* or *masculine* can deliver different mental images in Japan and other countries and 5 years ago versus today.

Kansei/affective engineering was founded by Mitsuo Nagamachi at Hiroshima University about 35 years ago (Nagamachi, 1989; 1995; 1999; 2007). Kansei/affective engineering aims at the discovery and translation of the customer's affective and emotional needs and is considered one of the best-structured methodologies in the world to translate the Kansei needs into attributes of the new product, its characteristics, and its functions (i.e., design elements or decision rules). It is also known by the names of affective engineering, affective ergonomics, and emotional engineering; and it is an evolving (i.e., enriched through other tools and methods) customer-oriented product development technology that invariably leads to a better satisfaction as products and services match and/or exceed customers' profound needs, desires, and feelings (Hirata, 2008).

Different types of Kansei/affective engineering have been defined by Dr. Nagamachi, and they depend on successful applications in new fields based on the following elements: Totally new product or innovation from an existing product; clear definition of customer needs (explicit needs); clear definition of technical characteristics that make customer satisfaction possible; computer system databases with a knowledge base that controls the system and modeling; virtual imaging; methodological approach to define design rules; and finally, product, service, or community-development orientation.

The customer's psychological responses (Kansei) are more general qualitative characteristics, and in consequence, difficult to measure; but in order to transfer Kansei into design elements, qualitative psychological phenomena should be changed to quantified characteristics (linking Kansei with design technical specifications). For example, in the beverage industry, if the developmental target is the design of a new beverage flavor, we must not only define the market's specific segment and its functional needs regarding quality, quantity, cost and delivery, safety, and service, but also understand its Kansei needs (e.g., "I want an urbanlike drink," or "I want a attractive drink"), translate these profound voices of the customer, and deploy them into technical characteristics or design elements that satisfy them (e.g., can color, brand name, letter font, percentage of CO_2, dryness, liquid color).

The applications of Kansei/affective engineering can be found in the automotive industry, construction machinery, electric home appliances, office automation machinery, audiovisual equipment, home construction materials, shoe and garment industry, cosmetic industries and laboratories, stationery products, community design projects, and food industry.

It is important to note that in highly competitive markets, the Kansei/affective engineering approach has a purpose to *enhance quality of life through customer satisfaction.*

The general model of Kansei/affective engineering has the following phases:

1. Selection of the product or service domain and definition of the strategy, which includes the selection of the product or service (existing or totally new one), definition of the market and current

competition with its solutions, potential market segments, senses to be used in the study (sight, taste, smell, touch, hearing) and their combinations, and finally, the general definition of the strategy and plan of the project. This phase shall include potential concepts as well as solutions not yet developed in order to cover a larger scope of the domain.

2. Definition of the semantic space and its structure, which include the collection of adjectives that describe the product or service domain and the potential Kansei needs (i.e., profound needs of the market) called Kansei words (e.g., elegant, masculine, sober, attractive, urbanlike, sexy, heavy), their categorization, definition of the hierarchical structure, and data collection. The Kansei words are collected from various sources such as the team of experts, designers, experienced users, advertisements, magazines, ideas, direct observation, and interviews. The list can go from 50 to more than 500 Kansei words and is commonly categorized (i.e., in groups) in a manual or statistical approach. In the manual form, a group of experts hierarchically organize the Kansei words depending on how specific or general the adjective is (Figure 11.1 shows an example of the hierarchical structure of Kansei words). The statistical approaches include factor analysis, principal component analysis, cluster analysis, and others (Figure 11.2 shows an example of a principal component analysis graph to reduce the dimensionality of Kansei words). The main objective is to determine the most representative Kansei needs.

3. Definition of product or service properties. The objective is to determine potential properties or design elements (i.e., technical and design requirements) of the future product or service, which include the collection of existing products, creation of new concepts, identification of potential customer and company images and priorities, as well as the definition of properties, elements (i.e., attributes or characteristics) and design categories.

4. Data collection. This phase is where the semantic space or the Kansei needs are related to the potential product or service properties through evaluations made by user surveys, direct observation, or physiological measurements. Kansei need is a response variable, and the potential properties or design elements are the independent variables of the model.

5. Data analysis. Data are analyzed and can be processed through manual (e.g., category classification method), statistical (e.g., regression analysis), or nonstatistical methods (e.g., rough sets theory; Nishino, 2005) in order to obtain the best approximation for the relation between the profound Kansei needs and the design elements (Figure 11.3 shows an example of the translation of Kansei needs into design elements).

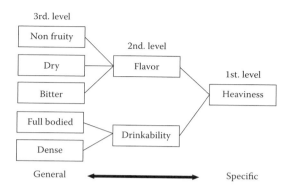

FIGURE 11.1
Hierarchical Kansei word structure.

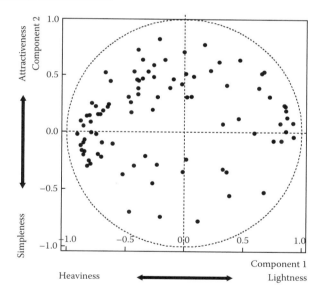

FIGURE 11.2
Principal component analysis of Kansei words (Hirata, R., Nagamachi, M., Ishihara, S. (2004a). Satisfying emotional needs of the beer consumer through Kansei Engineering Study, *Proceedings of the 7th International QMOD Conference 2004*, University of Linköping and ITESM; Hirata, R., Nagamachi, M., Ishihara, S. (2004b). Satisfying emotional needs of Mexican beer consumers market through Kansei Engineering Study, *Proceedings of 10th ICQFD Conference 2004*, Mexico.).

6. Validation and prototype construction. All the results shall be tested for validity and discussed with the team of experts (e.g., technicians, designers, expert users) before proceeding to the construction of prototypes and market testing, as well as planning for the next production and market introduction processes (i.e., development and commercialization phases).

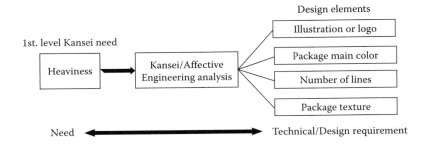

FIGURE 11.3
Kansei need translation to design elements.

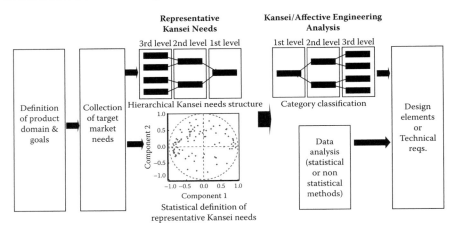

FIGURE 11.4
Translation of market needs and Kansei/affective engineering.

Kansei/affective engineering is a structured approach for the identification, categorization, and translation of profound Kansei needs of the market into physical design elements or technical requirements, using a diverse variety of quantitative and qualitative tools. Figure 11.4 shows a general process for Kansei/affective engineering.

11.4 Quality Function Deployment

Quality function deployment (QFD) was originated in Japan by Dr. Joji Akao while linking the critical points of quality assurance to be carried out through the design and manufacturing processes. Around 1972, he and Dr. Shigeru Mizuno developed a matrix of customer demands and quality characteristics that was used at the Kobe Shipyards of Mitsubishi Heavy Industries. Today, QFD is a comprehensive method for the product or service design and planning ensuring customer satisfaction through meeting and translating his requirements and demands throughout each stage of the product or service development process (Akao 1994; Akao and Mizuno 1994). Its main function is to translate customer needs and desires into the technical requirements of the product or service, as well as deploying its features, components, required technologies, process characteristics, required capability and reliability, and deliverable actions, among others. It links the demanded quality of the market with design, development, engineering, production, and service functions, aligning all company departments to the construction of the benefits the customer requirements and desires (i.e., value from the customer perspective) rather than elimination of errors and claims of defects.

The QFD process gathers and organizes the customer needs and then tailors a specific strategy in order to translate or deploy the market requirements into means to accomplish them using matrix relationships (e.g., target–means matrix) where rows represent the requirements and columns, the alternative or potential means. Figure 11.5 presents an example of how QFD can deploy demanded quality from the market into technical characteristics (quality deployment, matrix 1), demanded quality into product functions (function deployment, matrix 2), demanded quality into failures to avoid (reliability deployment, matrix 3), as well as further deployments such as the relation of functions and technology (technology deployment, matrix 4).

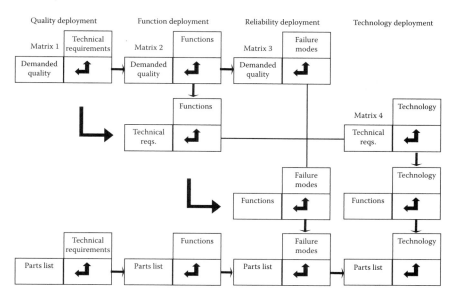

FIGURE 11.5
Comprehensive QFD (example of deployments).

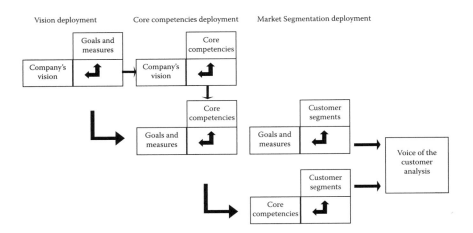

FIGURE 11.6
Vision deployments with QFD matrix (example of deployments).

As a planning and design process, a modern QFD matrix is also used to bridge the gap between the company's priorities and needs with product designs. QFD links to strategic planning deploying the company's vision into its business plan, strategy definition, market segmentation, and customer selection, where it connects into the traditional QFD with the definition of customer needs and demands (Shillito 1994). Figure 11.6 shows an example of the vision deployment into customer needs analysis.

One of the key matrices in the QFD methodology is the relationship matrix in the quality deployment stage (Figure 11.7, matrix 1), where customer needs are related to technical requirements by a group of experts and generally built as the first relationship for product planning. QFD requires the demanded quality to be prioritized, in order to determine key technical requirements and its targets. For this purpose a quality planning table (QPT) is constructed, which includes customer priorities, an evaluation between the company's current products and the competitive products and other elements such as sales opportunities (Terninko 1997). Other relationship matrices such as the product planning table (PPT) are added and create what is commonly known as the House of Quality (HoQ) (Figure 11.7 is an image of the HoQ with its matrices).

Customer needs are collected, interpreted, and organized through various methods such as focus groups, consumer brainstorming, nominal group technique, ethnography (Mariampolski 2006), voice of the customer (VOC) analysis (Katz 2004; Shillito 2001), in-depth interviews, customer visits (McQuarrie 1998), lead user analyses (Von Hippel 1988, 2005), and others. The key point is to clearly differentiate and separate the customer's expected features (functions) and the required and expected benefits, named

House of Quality (HoQ)

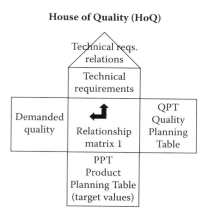

FIGURE 11.7
Quality deployments and the HoQ.

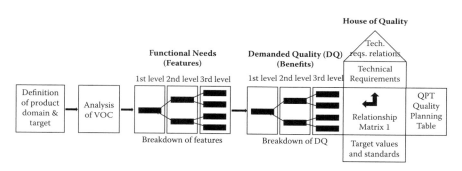

FIGURE 11.8
From VOC to DQ for the HoQ construction.

demanded quality (DQ) for the correct input of the relationship matrix and HoQ. Figure 11.8 shows a general process for the definition of the DQ.

11.5 QFD and Kansei

Since QFD for product planning generally focuses on technical solutions, Kansei needs are difficult to identify, collect, or analyze from the gathered data. Additionally, common tools for traditional QFD grasp explicit customer needs and link them to technical requirements and design elements known by a group of experts (i.e., explicit potential solutions, measures, features).

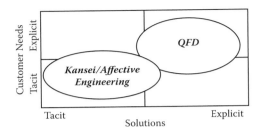

FIGURE 11.9
Relation between Kansei/affective engineering and QFD scope (Hirata, R. (2005). Understanding emotional needs of the Japanese and Mexican beer consumer market through Kansei Engineering Study, presented at International Congress on Quality, Tokyo, 2005.).

Kansei/affective engineering is very useful and an effective technology from the outset, when customer needs are profound and need to be discovered, gathered, and evaluated before deploying solutions (i.e., tacit or unconscious needs). Second, it translates the Kansei needs into technical requirements or design elements not necessarily known by the group of experts (i.e., tacit or unknown potential solutions, measures, features).

Once the Kansei needs and the technical requirements are placed in the explicit arena, QFD is a powerful and profitable technology for the deployment of the product or service. The relation between Kansei/affective engineering and QFD regarding its scope is shown in Figure 11.9 (Hirata, 2005).

Kansei/affective engineering is useful and can articulate with QFD in the identification and categorization of profound, tacit, emotional, and affective needs (i.e., Kansei needs) and is complementary (or prior) to the QFD (VOC) analysis. Both approaches will feed the relationship matrix of the HoQ with categorized and organized explicit demanded quality (DQ) elements (e.g., breakdown of DQ). In QFD, a group of experts defines the possible technical requirements for the DQ elements, and the strengths of their relationships are also determined by distinguishing between a strong, medium, or weak relationship, which accelerates the deployment and the decision making processes under the assumption that all the technical requirements are known, as well as necessary and sufficient. In case the assumption cannot be assured, then further quantitative studies shall be done.

Regarding the determination of the technical requirements or design elements for Kansei needs, Kansei/affective engineering offers quantitative evidence through statistical and nonstatistical approaches (e.g., regression analysis, rough sets) and as a consequence, it also provides a more detailed quantification of the relationships between the DQ and the technical requirements giving more certainty to the design or experts' team. Kansei/affective engineering also feeds the HoQ with the definition of technical requirements or physical design elements, which then can be deployed into other QFD functions. A full approach with QFD and Kansei/affective engineering is shown in Figure 11.10 (Hirata, 2005).

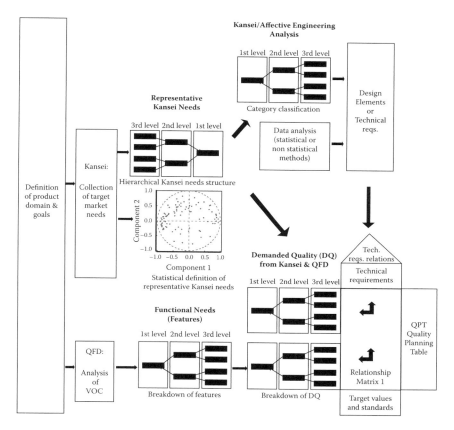

FIGURE 11.10
Kansei/affective engineering and QFD interaction toward HoQ (Hirata, R. (2005). Understanding emotional needs of the Japanese and Mexican beer consumer market through Kansei Engineering Study, presented at International Congress on Quality, Tokyo, 2005.).

11.6 Conclusions

Translation of affective and emotional needs of the markets into design elements and attributes of products and services is necessary to satisfy profound human needs. The discovery phase in the new product development process needs tools and methods to identify, organize, and translate Kansei needs into product or service specifications, and companies shall integrate Kansei/affective engineering as a method for this purpose. In product planning, Kansei/affective engineering is useful to identify the Kansei needs of the target market, especially in the current competitive environments where customer satisfaction is evolving from the satisfaction of the obvious and explicit needs, mostly functionality and usability demands, toward the satisfaction of profound, tacit, emotional, and affective needs.

Kansei/affective engineering, as well as QFD, is a flexible and evolving model that allows the addition and integration of other tools and methods throughout their various stages. Therefore, both can articulate in the process of product and service development providing the tools and processes to translate Kansei needs into design elements, as well as the numerical evidence to support the decision making processes bringing shorter product development cycle times, reduced development costs, and increased satisfaction of functional, usability, affective, and emotional customer needs and in consequence, a better success rate of new products and services in the market.

References

Altshuller, G. S. (1999). *The Innovation Algorithm*, Technical Innovation Center, Worcester, MA.

Akao, J. (1994). *Quality Function Deployment*, Productivity Press, New York.

Akao, J., and Mizuno, S. (1994). QFD: Customer Driven Approach to Quality Planning and Deployment. Asian Productivity Organization, Tokyo.

Cooper, R., and Edgett, S. (2008). Ideation for product innovation: What are the best methods? *Product Innovation Best Practices Series*, Reference paper #29, Stage-Gate Inc. & Product Development Institute.

Green, P., Wind, J., and Rao, V. R. (1999). Conjoint analysis: Methods and applications. In Dorf, R. C., (Ed.), *The Technology Management Handbook*, CRC Press, New York.

Garcia, M. L., and Bray, O. H. (1997). *Fundamentals of Technology Roadmapping*, Sandia National Laboratories, Albuquerque, NM.

Green, W. S., and Jordan, P. W. (Eds.) (2002). *Pleasurable Products: Beyond Usability*, CRC Press, New York.

Griffith, A., and Belliveau, P. (1997). *Drivers of NPD Success: The 1997 PDMA Report*, Product Development & Management Association PDMA, Mount Laurel, NJ.

Hirata, R. (2005). Understanding emotional needs of the Japanese and Mexican beer consumer market through Kansei Engineering Study, presented at International Congress on Quality, Tokyo, 2005.

Hirata, R. (2009). Traducción de las emociones y sensaciones del cliente en productos y servicios, Ph.D. Dissertation, UNAM, Graduate School of Management Science, Mexico City, Mexico.

Hirata, R., Nagamachi, M., Ishihara, S. (2004a). Satisfying emotional needs of the beer consumer through Kansei Engineering Study, *Proceedings of the 7th International QMOD Conference 2004*, University of Linköping and ITESM.

Hirata, R., Nagamachi, M., Ishihara, S. (2004b). Satisfying emotional needs of Mexican beer consumers market through Kansei Engineering Study, *Proceedings of 10th ICQFD Conference 2004*, Mexico.

Hirata, R., Nagamachi, M., Ishihara, S., Nishino, T. (2008). Translation of customer Kansei and emotional needs into products, *The 2nd International Conference on Applied Human Factors and Ergonomics* (AHFEI) 2008, Las Vegas.

Jordan, P. W. (2000). *Designing Pleasurable Products*, CRC Press, London.

Kahaner, L. (1998). *Competitive Intelligence*, Touchstone, New York, and European University Association, Brussels.

Kahn, E. (2003). Patent mining in a changing world of technology and product development, *Intellectual Asset Management*, July/August.

Katz, G. M. (2004). The vVoice of the customer, Chapter 7. In P. Belliveau, A. Griffin, and S. Somermeyer (Eds.), *PDMA Toolbook 2 for New Product Development*, John Wiley & Sons, Hoboken, NJ.

Katz, G. (2007). The PDMA's body of knowledge. In *PDMA Toolbook 3 for New Product Development*, Griffin, A., and Somermeyer, S. (Eds.), John Wiley & Sons, Hoboken, NJ.

Mariampolski, H. (2006). *Ethnography for Marketers: A Guide to Consumer Immersion*, Sage Publications, Thousand Oaks, CA.

McQuarrie, E. F. (1998). *Customer Visits*, Sage Publications, European University Association, Brussels.

Nagamachi, M. (1989). *Kansei Engineering*, Kaibundo Publishing, Japan.

Nagamachi, M. (1995). *The Story of Kansei Engineering*, Japanese Standards Association, Tokyo.

Nagamachi, M. (1999). Kansei engineering: A new consumer oriented technology for product development. In W. Karwowski and W. S. Morris (Eds.), *The Occupational Ergonomics Handbook*, CRC Press, New York.

Nagamachi, M. (2004). Kansei engineering. In N. Stanton and A. Hedge et al. (Eds.), *Handbook of Human Factors and Ergonomics Methods*, CRC Press, New York.

Nagamachi, M. (2007). Perspectives and new trend of Kansei/affective engineering, *The 1st European Conference on Affective Design and Kansei Engineering & 10th QMOD Conference*, University of Linkoping and Lund University, Helsingborg.

Nagamachi, M., and Imada, A. (1995). Kansei engineering: An ergonomic technology for product development, *International Journal of Industrial Ergonomics*, **15** (1), January.

Nishino, T. (2005). Rough sets and Kansei rules definition. In M. Nagamachi (Ed.), *Product Development and Kansei*, Chapter 9, Kaibundo, Tokyo.

PDMA Foundation (2004). Comparative Performance Assessment Study, PDMA, USA.

PDMA (2009). Product Development and Management Association, Body of Knowledge, http://pdmabok.arcstone.com/(accessed September 30, 2009).

Schifferstein, H. N. J., and Hekkert, P. (Eds.) (2008). *Product Experience*, Elsevier, Amsterdam.

Shillito, L. (1994). *Advanced QFD*, John Wiley & Sons, New York.

Shillito, L. (2001). *Acquiring, Processing and Deploying Voice of the Customer*, CRC Press, New York.

Terninko, J. (1997). *Step by Step QFD*, 2nd ed., St. Lucie Press, Boca Raton, FL.

Von Hippel, E. (1988). *The Sources of Innovation*, Oxford University Press, Oxford, U.K.

Von Hippel, E. (2005). *Democratizing Innovation*, The MIT Press, Cambridge, MA.

Index